Biochemistry:
A Synopsis

A Concise Medical Library for Practitioner and Student

Biochemistry:
A Synopsis

Diane S. Colby, PhD

Adjunct Assistant Professor
Department of Biochemistry and Biophysics
University of California School of Medicine
San Francisco

LANGE Medical Publications
Los Altos, California 94023

Table of Contents

SECTION III. CELL BIOLOGY

Preface

This *Synopsis* is intended to serve as a review of biochemistry for students currently taking courses and others who require review in preparation for examinations such as the National Board or FLEX examinations. Because of increasing demands on students' time, there is a need to present information in a concise and efficient manner. Thus, this book is suitable also as a "short-course" text for self-study or for students in related fields who need basic knowledge of the subject. It covers all of the topics conventionally included in survey courses offered to health professionals. Each chapter provides a succinct summary of the basic concepts of the subject area and sets forth objectives designed to facilitate efficient study. The sample examination questions are modeled after those asked on various certifying and Board examinations.

To assist the student in both learning and retaining the material, I have endeavored to explain the physiologic significance of each of the topics presented. The student is urged to ponder the objectives both before and after reading the text of each chapter—before, to focus attention on the important topics; and after, to make certain the important points have been understood. The sample questions should be used as reinforcement. In order to simulate the conditions of the National Board examination, the questions are grouped not by chapter but in three sections that parallel the organization of the book.

I wish to express my appreciation to Gil Fernandes for his skillful rendering of the original art.

Diane S. Colby, PhD

San Francisco
October, 1985

Section I:
Protein Structure & Function

A Review of Chemistry | 1

OBJECTIVES

- Given the $\Delta G^{0'}$ for a reaction, be able to calculate the K_{eq}.

- Given either the K_{eq} or the $\Delta G^{0'}$ for a reaction, be able to predict the equilibrium concentrations of the reactants and products.

- Be able to predict whether a reaction will proceed spontaneously given either the $\Delta G^{0'}$ and the initial concentrations of reactants and products or the ΔG for the reaction.

- Be able to calculate the $\Delta G^{0'}$ for a series of reactions given the $\Delta G^{0'}$ of each reaction in the series.

◄ • ►

BECAUSE AN understanding of the basic principles and conventions of chemistry is a prerequisite to the study of biochemistry, this book begins by reviewing those aspects of chemistry with which the reader is expected to be familiar.

Table 1–1. Types of chemical bonds. Covalent bonds are indicated by solid lines between atoms; noncovalent bonds are indicated by dotted lines.

Type of Bond	Example	Energy Required to Break Bond
Covalent A covalent bond is formed when one atom shares its electrons with another atom.	 Methane, which has 4 covalent bonds.	20–200 kcal/mol
Noncovalent An ionic bond is formed by the attraction of 2 groups of opposite charge.	 An ionic bond between a carboxylic acid group and an amino group.	10–100 kcal/mol
A hydrogen (H) bond is formed when the proton of a hydrogen atom that is covalently bonded to one electronegative atom is attracted to a second electronegative atom.	 An H bond between a secondary amino group and a ketone.	5 kcal/mol
Van der Waals attractions occur between atoms that are in contact with each other. The attraction results from the electrostatic attraction of the positive nucleus of one atom for the negative electron cloud of its neighbor.	 Multiple van der Waals attractions between 2 isoleucine side chains.	0.5 kcal/mol

Chemical Bonds

As you read this book, it will be important to keep in mind the definitions of several types of chemical bonds (summarized in Table 1–1). A **covalent bond** is one in which electrons are shared between 2 atoms that have incomplete outer electron shells. A group of atoms held together by covalent bonds is termed a **molecule.** Covalent bonds are strong; ie, a large amount of energy is required to break them. **Noncovalent bonds consist of electrostatic attractions (the attraction of opposite charges) between atoms that have complete outer electron shells.** Noncovalent bonds, which are relatively weak, determine how molecules associate with each other. The 3 types of noncovalent bonds most important in biochemistry are **ionic bonds, hydrogen bonds** (abbreviated H bonds), and **van der Waals attractions.** An ionic bond is the bond formed by the attraction of 2 groups that carry opposite charges. Charged groups of importance in biology include amino, carboxyl, and phosphoryl groups. An H bond is formed when a proton that is covalently bonded to one electronegative atom is attracted to a second electronegative atom. In biologic systems, O and N atoms are often involved in H bonds. Van der Waals attractions result from the electrostatic attraction of the positive nucleus of one atom for the negative electron cloud of a neighboring atom. These are the weakest of the attractive forces, and they are effective only over short distances. As a consequence, van der Waals attractions are effective in holding 2 molecules together only when several atoms of each molecule are in contact with each other. Multiple close contacts are formed between molecules that have surfaces which are complementary in shape.

Functional Groups

Many biologic molecules are extremely complex structures. The study of them can be simplified, however, if they are viewed as made up of recognizable smaller units called functional groups. This way of approaching complex compounds is useful for 2 reasons. (1) Because the properties of the functional groups are retained in the larger compounds of which they are a part, it is often possible to predict the behavior of a complex molecule from the behavior of its functional groups. (2) The conversion of one biologic molecule to another usually involves alteration of only one functional group in any given reaction. This principle is illustrated in the 2-step conversion of glycerol to dihydroxyacetone phosphate.

| Glycerol | Glycerol 3-phosphate | Dihydroxyacetone phosphate |

Table 1–2 lists the functional groups most commonly encountered in biochemistry and the reactions in which they participate.

Table 1–2. Functional groups important in biochemistry.

Functional Group	Reactions
Alcohol Primary $R—CH_2—OH$	(1) Oxidation $R—CH_2—OH \xrightarrow[\text{H}_2\text{O}]{[O]} R—\overset{\displaystyle O}{\underset{\displaystyle}{C}}—H$ A primary alcohol An aldehyde
Secondary $\underset{R_2}{\overset{R_1}{HC}}—OH$	(1) Oxidation $\underset{R_2}{\overset{R_1}{HC}}—OH \xrightarrow[\text{H}_2\text{O}]{[O]} \underset{R_2}{\overset{R_1}{C}}=O$ A secondary alcohol A ketone (2) Esterification $R_1—CH_2—OH + HO—\overset{\displaystyle O}{C}—R_2 \xrightarrow[\text{H}_2\text{O}]{} R_1—CH_2—O—\overset{\displaystyle O}{C}—R_2$ A primary alcohol An acid An ester

Aldehyde
 	
$R—\overset{\displaystyle O}{C}—H$
	(1) Oxidation
	
	$R—\overset{\displaystyle O}{C}—H \xrightarrow{[O]} R—\overset{\displaystyle O}{C}—OH$
	          An acid
	
	(2) Reduction
	
	$R—\overset{\displaystyle O}{C}—H \xrightarrow{[2H]} R—CH_2—OH$
	          An alcohol
	
	(3) Acetal or hemiacetal formation
	
	$R_1—\overset{\displaystyle O}{C}—H + R_2OH \longrightarrow R_1—\overset{\displaystyle H}{\underset{\displaystyle \underset{R_2}{O}}{C}}—OH$
	
	                    A hemiacetal
	
	$R_1—\overset{\displaystyle O}{C}—H + 2\,R_2OH \xrightarrow[\text{H}_2\text{O}]{} R_1—\overset{\displaystyle H}{\underset{\displaystyle \underset{R_2}{O}}{C}}—O—R_2$
	
	                    An acetal
	
	(4) Aldol condensation
	
	$R—\overset{\displaystyle H}{C}=O + CH_3—\overset{\displaystyle H}{C}=O \longrightarrow R—\underset{\displaystyle OH}{\overset{\displaystyle H}{C}}—CH_2—\overset{\displaystyle H}{C}=O$

Table 1–2 (cont'd). Functional groups important in biochemistry.

Functional Group	Reactions
Ketone $\begin{array}{c} R_1 \\ \diagdown \\ C=O \\ \diagup \\ R_2 \end{array}$	**(1) Reduction** $\begin{array}{c} R_1 \\ \diagdown \\ C=O \\ \diagup \\ R_2 \end{array} \xrightarrow{[2H]} \begin{array}{c} R_1 \\ \diagdown \\ HC-OH \\ \diagup \\ R_2 \end{array}$ An alcohol
Carboxylic acid $\begin{array}{c} O \\ \parallel \\ R-C-OH \end{array}$	**(1) Reduction** $\begin{array}{c} O \\ \parallel \\ R-C-OH \end{array} \xrightarrow{[2H]} \begin{array}{c} O \\ \parallel \\ R-C-H \end{array}$ $\searrow H_2O$ An aldehyde **(2) Esterification (see alcohols)** **(3) Acid anhydride formation** $\begin{array}{c} O \\ \parallel \\ R_1-C-OH \end{array} + \begin{array}{c} O \\ \parallel \\ HO-C-R_2 \end{array} \longrightarrow \begin{array}{c} O \quad O \\ \parallel \quad \parallel \\ R_1-C-O-C-R_2 \end{array}$ $\searrow H_2O$ **(4) Amide formation** $\begin{array}{c} O \\ \parallel \\ R_1-C-OH \end{array} + R_2-NH_2 \longrightarrow \begin{array}{c} O \\ \parallel \\ R_1-C-NH-R_2 \end{array}$ $\searrow H_2O$ An amine An amide **(5) Salt formation** $R-C\begin{array}{c} O^- \\ \diagdown \\ \diagup \\ O \end{array} + Na^+ \longrightarrow R-C\begin{array}{c} O^-\cdots Na^+ \\ \diagdown \\ \diagup \\ O \end{array}$
Amine Primary $R-NH_2$ Secondary $\begin{array}{c} R_1 \\ \diagdown \\ NH \\ \diagup \\ R_2 \end{array}$ Tertiary $\begin{array}{c} R_2 \\ \mid \\ R_1-N-R_3 \end{array}$	**(1)** In solution, the charged and uncharged forms of an amine are in equilibrium. $R-NH_2 \rightleftharpoons R-NH_3^+$ $+$ H^+

Table 1–2 (cont'd). Functional groups important in biochemistry.

Functional Group	Reactions
Phosphate Phosphate ester $$R-O-\overset{\overset{\displaystyle O}{\|}}{\underset{\underset{\displaystyle OH}{\|}}{P}}-OH$$	(1) Phosphoryl transfer $$R_1-O-\overset{\overset{\displaystyle O}{\|}}{\underset{\underset{\displaystyle OH}{\|}}{P}}-OH + R_2OH \longrightarrow R_1OH + R_2-O-\overset{\overset{\displaystyle O}{\|}}{\underset{\underset{\displaystyle OH}{\|}}{P}}-OH$$
Phosphodiester $$R_1-O-\overset{\overset{\displaystyle O}{\|}}{\underset{\underset{\displaystyle OH}{\|}}{P}}-O-R_2$$	
Phosphoanhydride $$R-O-\overset{\overset{\displaystyle O}{\|}}{\underset{\underset{\displaystyle OH}{\|}}{P}}-O-\overset{\overset{\displaystyle O}{\|}}{\underset{\underset{\displaystyle OH}{\|}}{P}}-OH$$	
Thiol (sulfhydryl) Thiol R—SH Thioester $$R_1-S-\overset{\overset{\displaystyle O}{\|}}{C}-R_2$$	(1) Disulfide formation $$R_1-SH + HS-R_2 \xrightarrow{\;[O]\;} R_1-S-S-R_2$$ H_2O

Stereoisomers

Most of the compounds discussed in this book contain carbon. Carbon plays a central role in biochemistry because it forms strong single and double carbon-carbon bonds ($C-C$ and $C=C$, respectively) and because it forms stable compounds with a variety of other elements.

* Carbon is tetravalent; ie, it must form 4 covalent bonds to complete its outer electron shell. Because each carbon can form several bonds, carbon-containing compounds often exhibit a type of isomerism called **stereoisomerism.** Two different compounds are isomers if they have the same molecular formula. Stereoisomers are isomers that are identical except for the order in which substituent groups are arranged around the carbon atom. Two types of stereoisomers are commonly encountered in biochemistry: mirror image isomers (**enantiomers**) and *cis-trans* isomers (**geometric isomers**).

When carbon forms single bonds to 4 groups, the bonded groups are arranged at the vertices of a tetrahedron, as for example in methane (Fig

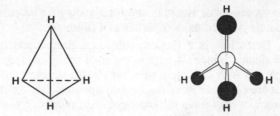

Figure 1–1. Tetrahedral and ball-and-stick model representations of methane (CH_4).

1–1). If the 4 bonded groups are all different, the central carbon is designated an **asymmetric** or a **chiral** carbon, and the compound can be drawn in 2 ways that are mirror images of each other (Fig 1–2). Two compounds are mirror image isomers if they cannot be superimposed upon each other regardless of how they are rotated in space.

The nomenclature of mirror image isomers was historically based on the observation that some pairs of isomers rotate the plane of plane-polarized light. One member of the pair (the dextrorotatory, or D, isomer) rotates plane-polarized light to the right, while the other (the levorotatory, or L, isomer) rotates it an equal amount to the left. The letters D and L are used today to designate not the direction of rotation of light but the absolute configuration of the substituent groups around the tetrahedral carbon atom. Glyceraldehyde serves as the standard for the description of all other mirror image isomers. The D isomer of any compound is that which can be synthesized from D-glyceraldehyde via reactions that do not alter the configuration around the asymmetric carbon.

Except for the direction in which they rotate plane-polarized light, members of a pair of mirror image isomers have identical physical properties. They also have identical chemical properties except when reacting with other compounds that are also mirror image isomers. Generally, only one member of a pair of enantiomers is biologically active. For example, only the D

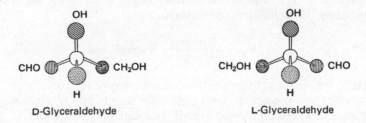

D-Glyceraldehyde L-Glyceraldehyde

Figure 1–2. Ball-and-stick model representations of glyceraldehyde enantiomers.

isomer of glucose can serve as a substrate for glycolysis, and only L isomers of amino acids are used in the biosynthesis of proteins.

Cis-trans isomerism is a property of some compounds that contain carbon-carbon double bonds. If both carbons of the double bond are also joined by single bonds to 2 different substituents, the structure may be drawn in 2 forms that are *cis-trans* isomers. Two compounds that constitute a pair of *cis-trans* isomers may have different common names, eg, maleate and fumarate.

$$
\begin{array}{cc}
\mathrm{H-\overset{\|}{C}-COO^-} & \mathrm{{}^-OOC-\overset{\|}{C}-H} \\
\mathrm{H-\overset{\|}{C}-COO^-} & \mathrm{H-\overset{\|}{C}-COO^-} \\
\text{Maleate \textit{(cis)}} & \text{Fumarate \textit{(trans)}}
\end{array}
$$

The designation *cis* indicates that the 2 larger substituents are on the same side of the double bond. If they are on opposite sides of the double bond, the compound is the *trans* isomer. *Cis-trans* isomers do not have identical physical or chemical properties.

Equilibrium Constants

Every reaction is, to some extent, reversible; ie, both the forward and reverse reactions take place. Often, it is important to describe which direction is favored. The following example illustrates the use of equilibrium constants for this purpose.

The enzyme phosphotriose isomerase catalyzes the interconversion of 2 compounds, glyceraldehyde 3-phosphate and dihydroxyacetone phosphate.

$$
\begin{array}{cc}
\mathrm{CH_2-PO_4{}^{2-}} & \mathrm{CH_2-PO_4{}^{2-}} \\
\mathrm{H-\overset{|}{C}-OH} & \mathrm{\overset{|}{C}{=}O} \\
\mathrm{H-\overset{|}{C}{=}O} & \mathrm{CH_2OH}
\end{array}
$$

Glyceraldehyde 3-phosphate $\rightleftharpoons$ Dihydroxyacetone phosphate

In any reaction, the compounds written to the left of the arrows are considered the reactants and those to the right the products.

If a reaction is allowed to proceed for a long enough time, it will reach equilibrium, after which time there will be no *net* change in the concentrations of compounds. The equilibrium distribution of reactants and products is described by the equilibrium constant (K_{eq}).

$$
K_{eq} = \frac{[\text{Products}]}{[\text{Reactants}]} \text{ at equilibrium}
$$

Brackets, [ ], are used to denote the concentration of the compound.

For any given reaction, the equilibrium distribution of reactants and products will be the same regardless of their initial concentrations; ie, each reaction has a characteristic K_{eq}. If, for example, the interconversion of glyceraldehyde 3-phosphate and dihydroxyacetone phosphate is carried out at pH 7.0, at 1 atmosphere pressure, and at 25 °C, at equilibrium 95% of the material will be dihydroxyacetone phosphate and 5% glyceraldehyde 3-phosphate. Thus,

$$K_{eq} = \frac{\text{[Dihydroxyacetone phosphate]}}{\text{[Glyceraldehyde 3-phosphate]}} = 20$$

The units of concentration are moles per liter, and thus, in this case, K_{eq} is unitless.

The dissociation constant (K_d) is used to express the extent to which a complex dissociates into its component parts. For example, acetic acid, CH_3COOH, dissociates to an anion, CH_3COO^-, and a proton, H^+. The K_d for this reaction is

$$K_d = \frac{[H^+][CH_3COO^-]}{[CH_3COOH]} = 1.76 \times 10^{-5} \text{ mol/L}$$

Free Energy Changes

✦A reaction is termed spontaneous if at equilibrium the concentration of the products is greater than that of the reactants. Based on the K_{eq} presented in the preceding section, we can see that the conversion of glyceraldehyde 3-phosphate to dihydroxyacetone phosphate is spontaneous.

Whether or not a particular reaction proceeds spontaneously depends on the **change in free energy** (ΔG) that occurs during the reaction. Free energy is that portion of the energy of a system that is available to do work.

All reactions proceed in the direction of lower free energy; ie, the final state of a system has less free energy than the initial state. Therefore, if we know the difference in free energy between the reactants and the products of a reaction, we can predict whether or not it will proceed spontaneously.

If ΔG is negative, the reaction proceeds spontaneously.

If $\Delta G = 0$, the reaction is at equilibrium.

If ΔG is positive, the reaction will not proceed spontaneously (and the reverse reaction will occur). Note that the term "spontaneous" pertains to an equilibrium property of a reaction and says nothing about the rate at which it will occur. *The rate of a reaction cannot be predicted from its ΔG value*.

ΔG has 2 components, or driving forces: ΔH, the change in enthalpy (internal energy or heat); and ΔS, the change in entropy (order).

$$\Delta G = \Delta H - T\Delta S$$

where T is the absolute temperature (in degrees Kelvin). As can be seen from this equation, a reaction can have a negative ΔG, and therefore proceed spontaneously, if it gives off heat (a negative ΔH) or if the molecules involved become more disordered or more numerous (a positive ΔS).

To simplify comparisons, the free energy change of a biologic reaction is generally reported as the standard free energy change ($\Delta G^{0'}$). $\Delta G^{0'}$ is defined as the value of ΔG (in kilocalories per mole) for a reaction conducted at pH 7.0, at 25 °C, and at 1 atmosphere pressure, and all reactants and products at concentrations of 1 mol/L (standard conditions for biologic reactions). The standard free energy change for any reaction is related to its K_{eq} by the following equation:

$$\Delta G^{0'} = - RT \ln K_{eq}$$

where R is the gas constant—approximately 2×10^{-3} kcal/degree/mol. The equation can be converted from ln to $\log_{10}$ by multiplying the right-hand half of the expression by 2.3. Applying the above relationship, we find that the $\Delta G^{0'}$ for the conversion of glyceraldehyde 3-phosphate to dihydroxyacetone phosphate is approximately -1.8 kcal/mol.

$$\Delta G^{0'} = - (2)(298)(2.3)(\log_{10} 20) = - 1.8 \text{ kcal/mol}$$

Of course, the conditions found in living organisms (in vivo) do not correspond to standard conditions, the most important difference being the concentrations of reactants and products. The free energy change that occurs under nonstandard conditions is related to $\Delta G^{0'}$ by the following equation:

$$\Delta G = \Delta G^{0'} + RT \ln \frac{[\text{Products}]}{[\text{Reactants}]}$$

Using this equation, we can predict the direction in which a reaction will proceed in vivo. Note that a reaction which has a positive (unfavorable) $\Delta G^{0'}$ may have a negative (favorable) ΔG if the ratio of products to reactants is very small. This means that in vivo a somewhat unfavorable reaction can occur if the reactants are present in high concentrations or if the products are continuously removed.

Although all reactions are reversible, in many cases either the forward or reverse reaction strongly predominates and the reaction is, in practical terms, irreversible. A readily reversible reaction has a small numerical value of $\Delta G^{0'}$, eg, $\pm$ 1 kcal/mol. One with a large negative value of $\Delta G^{0'}$ ($\Delta G^{0'}$ -5 kcal/mol) is effectively irreversible.

$\Delta G^{0'}$ values are additive. We can calculate the free energy change for a sequence of reactions that forms a pathway by adding the free energy changes for the individual steps. For example, the $\Delta G^{0'}$ for the conversion of A to D in the following pathway is -4.4 kcal/mol.

A ⟶ B ⟶ C ⟶ D
$-$ 4 kcal/mol   + 0.3 kcal/mol $-$ 0.7 kcal/mol

Water | 2

OBJECTIVES

- Given the pK for a weak acid and the pH of a solution containing that acid, be able to calculate the extent of dissociation of the acid.

- Given the amounts of a weak acid and its conjugate base in a solution and the pK value of the acid, be able to calculate the pH of the solution.

- Given the titration curve for a weak acid, be able to identify the pK value (or values) of the acid and to identify the range of pH in which it can function as a buffer.

WATER CONSTITUTES 50–90% of the mass of animal tissues. Consequently, most biochemical molecules are found in an aqueous environment. The behavior of these molecules is affected by 2 important properties of water: its polarity and its tendency to dissociate.

Hydrogen-Bonding Properties of Water

All molecules can be classified as either **polar,** ie, those that have a net charge or an asymmetric distribution of charge; or **nonpolar,** ie, those that do not. Although the water molecule as a whole carries no net charge, its electron cloud is not uniformly distributed throughout the molecule. Instead, the electron density is greatest near the oxygen atom. As a result, the oxygen atom carries a partial negative charge (designated δ^-), and each hydrogen atom carries a partial positive charge (designated δ^+) (Fig 2–1).

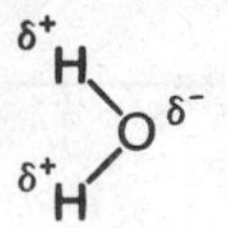

δ⁺
H
    O δ⁻
δ⁺
H

Figure 2–1. Water is a dipole. Owing to the uneven distribution of the electron cloud within the water molecule, the oxygen atom bears a partial negative charge (δ^-), and each of the hydrogen atoms bears a partial positive charge (δ^+).

A molecule in which the distribution of electrons creates a positive end and a negative end is termed a **dipole.** The dipolar character of water allows it to form H bonds with other water molecules and other polar substances (Fig 2–2).

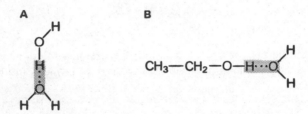

Figure 2–2. Water forms H bonds with itself and other polar compounds. *A:* Association of 2 water molecules. *B:* Formation of a hydrogen bond between water and ethanol. Hydrogen bonds are indicated by dotted lines.

The polarity of the water molecule is important for the organization of biochemical molecules. Polar substances dissolve in water easily because they can form H bonds with the polar water molecules. Nonpolar molecules are not easily dissolved in water for 2 reasons: They do not themselves form H bonds with water, and their presence in water prevents the water molecules from forming the maximum number of H bonds with each other. Polar substances are also designated __hydrophilic (water-loving)__ and nonpolar substances **hydrophobic** (water-fearing). When a nonpolar substance is added to water, it tends to form a separate, hydrophobic phase. (For a biologic application of this principle, see p 22.)

Dissociation of Water

On occasion, a proton that is involved in a hydrogen bond between 2 water molecules is transferred from the oxygen to which it was covalently bonded to that with which it was shared. The result of this transfer is the dissociation of a water molecule to form a hydroxyl (OH^-) ion and a hydronium (H_3O^+) ion.

$$2\,H_2O \rightleftharpoons H_3O^+ + OH^-$$

By convention, a hydronium ion is represented simply as H^+.

The extent to which water dissociates is very limited. In pure water,

$$[H_2O] = 55.56 \text{ mol/L}$$
$$[H^+] = 10^{-7} \text{ mol/L}$$
$$[OH^-] = 10^{-7} \text{ mol/L}$$

The K_d of water is thus

$$K_d = \frac{[H^+][OH^-]}{[H_2O]} = 1.8 \times 10^{-16} \text{ mol/L}$$

Because the dissociation of water is very limited, it does not significantly affect the concentration of molecular water. If we make the assumption that $[H_2O]$ is essentially constant, then the product $[H^+][OH^-]$, called the ion product for water, K_w, is also constant and equal to 10^{-14} (mol/L)^2. It then follows that the concentrations of H^+ and OH^- in solution are in reciprocal relationship to each other: When $[H^+]$ increases, $[OH^-]$ decreases. If the concentration of one ion is measured, the other can be calculated.

It is common practice to describe solutions with respect to $[H^+]$, reported as pH. The pH of a solution is defined as the log of the reciprocal of the hydrogen ion concentration.

$$pH = \log \frac{1}{[H^+]} = -\log [H^+]$$

For example, the pH of a solution that contains 10^{-3} mol/L of H^+ is 3. The $[OH^-]$ in the same solution is 10^{-11}. Tenfold differences in $[H^+]$ are represented by pH values that differ by 1.

Henderson-Hasselbalch Equation

An acid is an electrolyte that dissociates in water, releasing protons,

$$HA \rightleftharpoons H^+ + A^-$$

The protonated form of the electrolyte (HA) is referred to as the acid and the unprotonated form (A^-) its conjugate base. Thus, an acid is a proton donor and a base a proton acceptor.

Acids differ in the extent to which they dissociate in water. Strong acids dissociate completely; weak acids do not. The probability that an acid will dissociate is indicated by its dissociation constant, K_d.

$$K_d = \frac{[H^+][A^-]}{[HA]}$$

A weak acid is one that has a K_d substantially less than 1.

Weak acids are biologically important for 2 reasons. (1) The dissociation of a weak acid affects the pH of the solution in which it is dissolved. This property of weak acids is used to establish and maintain the normal body pH. (2) The extent to which a weak acid dissociates is affected by the pH in its environment. Thus, the extent of dissociation of an acidic group that is part of a biologic molecule, and consequently its chemical properties, changes in response to fluctuations in pH.

Because the protonated and unprotonated forms of an acid have different biologic properties, it is important to be able to predict the extent of dissociation of an acid at any pH. It is convenient to reexpress the K_d in a form analogous to pH, the pK.

$$pK = -\log K_d$$

We can then use the Henderson-Hasselbalch equation to predict the extent of dissociation of the acid.

$$pH = pK + \log \frac{[A^-]}{[HA]}$$

If we solve this equation for the case in which $A^- = HA$, we find that the pK corresponds to that pH at which the acid is 50% dissociated. An acid is 91% dissociated at a pH value 1 unit above the pK value and 91% associated at a pH value 1 unit below the pK value (Table 2–1). Thus, most of the change in the extent of dissociation takes place within 1 pH unit of the pK of the acid.

Table 2–1. The effect of pH on the dissociation of acetic acid.

pH	Percent CH_3CCO^-	Percent CH_3COOH
1.75	0.09	99.91
2.75	0.9	99.1
3.75	9	91
4.75	50	50
5.75	91	9
6.75	99.1	0.1
7.75	99.91	0.01

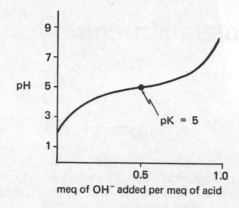

Figure 2–3. Titration curve for a weak acid having a pK= 5.

Buffers

Because the properties of many biologic molecules vary with pH, it is important that the pH of living systems be controlled within certain narrow limits. Biologic pH control is maintained by weak acids. When a solution containing a weak acid, such as acetic acid, is titrated with a strong base, such as NaOH, the increase in pH is not linearly related to the amount of base added (Fig 2–3). Instead, there is a plateau in the titration curve around the pH value that corresponds to the pK of the acid. This plateau is caused by the dissociation of the acid. When the solution is titrated with base, the concentration of free H^+ decreases. As the pH of the solution approaches the pK value of an acid, the acid begins to dissociate. The protons released in this process replace many of those titrated by the base. Thus, a solution that contains a weak acid resists changes in pH better than an equal volume of water. Because buffering depends on changes in the dissociation of a weak acid, a buffer is most effective within the range of pH where those changes are largest, ie, near the pK of the acid.

Fig 2–3 illustrates how titration data are used to determine the pK value of an acid. The pK corresponds to that point at which added base causes the least increase in pH, ie, the plateau in the titration curve.

tein Structure

OBJECTIVES

- Be able to recognize the structures of the 20 amino acids commonly found in proteins. Know the properties of their side chains and how these properties affect protein structure.

- Be able to recognize the structure of the peptide bond and to explain how its properties influence protein folding.

- Be able to describe the forces that determine how proteins fold.

- Know the definitions of the 4 levels of protein structure.

- Be able to describe the features of the major secondary structures of proteins.

- Know what is meant by protein denaturation.

ALL ORGANISMS make use of proteins to perform a number of functions that are essential for life. Among other things, proteins serve as catalysts, as molecular carriers, as receptors of biologic signals, and as structural components. Because proteins mediate a large number of important biologic processes, their properties determine many of the characteristics of living systems. In order to understand how they perform their essential functions, it is necessary first to understand their structural features.

L-α-Amino Acids

Proteins are polymers of L-α-amino acids. An amino acid is a compound that contains both an amino group and a carboxyl group (Fig 3–1). In an α-amino acid, both of these groups are attached to the same carbon atom,

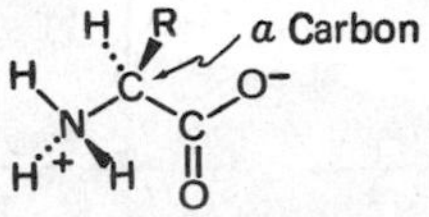

Figure 3–1. The structure of an L-α-amino acid. The R group is connected to the α carbon by a wedge to indicate that it is above the plane of the page. The α hydrogen is connected to the α carbon by a dotted line to indicate that it is below the plane of the page.

designated the α **carbon.** The α carbon of each amino acid is also bonded to an H atom and to a variable substituent, designated the R group or side chain. The configuration of the α-carbon substituents in an L-amino acid is as shown in Fig 3–1. When the amino group is drawn to the left, the α-carbon H lies behind the plane of the page and the R group in front of it. D-amino acids, in which the positions of the H and R group substituents are reversed, are not found in proteins.

The amino and carboxyl groups of an amino acid are able to ionize. The pK of the amino group is 9.8 ($\pm$ 1.0), and that of the carboxyl group is 2.1 ($\pm$ 0.5). Therefore, within the physiologic range of pH (pH 5.0–8.0), both groups are charged, as shown in Fig 3–1.

Table 3–1 lists the 20 common amino acids from which proteins are constructed. For convenience, they are often identified by the 3-letter abbreviations shown in the table. Nineteen of the amino acids differ only in the identity of the variable R group attached to the α carbon. The 20th, proline, is distinctive in that its side chain is a ring structure that includes the amino nitrogen.

Within Table 3–1, the amino acids are grouped according to the **polarity** of their variable R groups, a characteristic that strongly influences their behavior in proteins. Nine amino acids (glycine, alanine, valine, leucine, isoleucine, phenylalanine, methionine, proline, and tryptophan) have side chains that are nonpolar. These side chains have a limited solubility in water and tend to seek a hydrophobic environment. The remaining 11 amino acids have polar R groups. Within the range of physiologic pH, 5 (aspartate, glutamate, arginine, lysine, and histidine) are charged and 6 (serine, threonine, tyrosine, asparagine, glutamine, and cysteine) are uncharged. Table 3–1 lists the pK values of those side chains that are charged within the physiologic range of pH and shows that form of each side chain which predominates in vivo. With the exception of that of histidine, all of these pK values lie outside the physiologic range of pH. Thus, histidine is the only amino acid whose side chain undergoes changes in the extent of ionization under physiologic conditions. The exceptional behavior of histidine gives it a unique role in proteins as a buffer and as a physiologic switch (see Chapter 4).

Names, structures, and pK values of the 20 common amino acids.

Name (Abbreviation)	Structure*	pK of α-Carboxyl Group	pK of α-Amino Group	pK of R Group
Amino acids with nonpolar R groups				
Glycine (Gly)	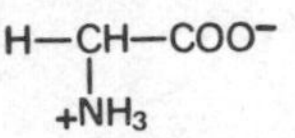 H—CH—COO⁻ ⁺NH₃	2.3	9.6	
Alanine (Ala)	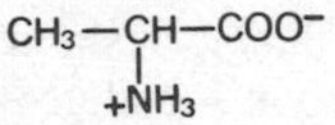 CH₃—CH—COO⁻ ⁺NH₃	2.3	9.7	
Valine (Val)	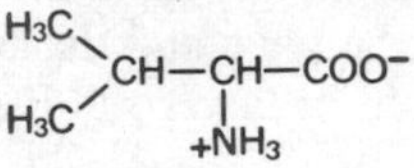 H₃C⧵CH—CH—COO⁻ H₃C⧸ ⁺NH₃	2.3	9.6	
Leucine (Leu)	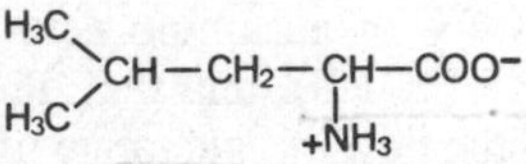 H₃C⧵CH—CH₂—CH—COO⁻ H₃C⧸ ⁺NH₃	2.4	9.6	
Isoleucine (Ile)	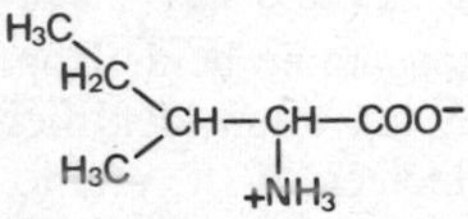 H₃C⧵H₂C⧵CH—CH—COO⁻ H₃C⧸ ⁺NH₃	2.4	9.7	
Phenylalanine (Phe)	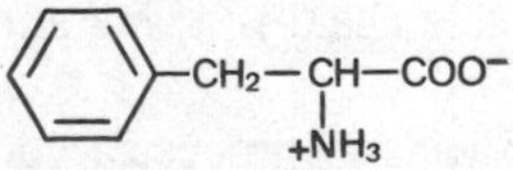 ⬡—CH₂—CH—COO⁻ ⁺NH₃	1.8	9.1	
Methionine (Met)	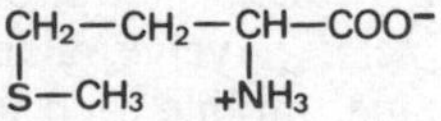 CH₂—CH₂—CH—COO⁻ S—CH₃ ⁺NH₃	2.3	9.2	
Proline (Pro)	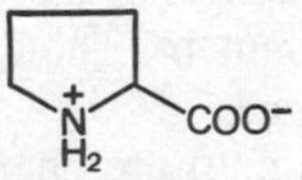 [pyrrolidine ring]—COO⁻ ⁺N H₂	2.0	10.6	
Tryptophan (Trp)	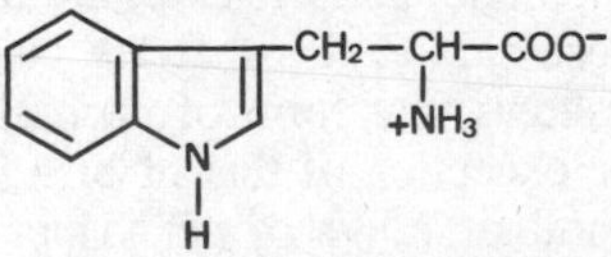 [indole]—CH₂—CH—COO⁻ ⁺NH₃ N H	2.4	9.4	
Amino acids with neutral polar R groups				
Serine (Ser)	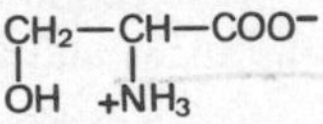 CH₂—CH—COO⁻ OH ⁺NH₃	2.2	9.2	

*The structure shown is that which predominates at pH 7.4.

Table 3–1 (cont'd). Names, structures, and pK values of the 20 common amino acids.

Name (Abbreviation)	Structure*	pK of α-Carboxyl Group	pK of α-Amino Group	pK of R Group
Amino acids with neutral polar R groups (cont'd)				
Threonine (Thr)	$CH_3-CH-CH-COO^-$ $\quad\quad OH \;\; +NH_3$	2.6	10.4	
Tyrosine (Tyr)	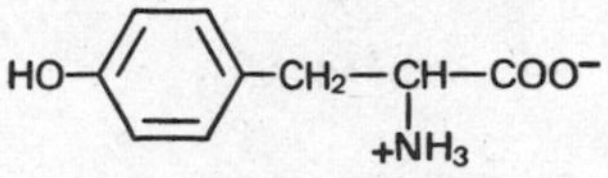 $HO-\bigcirc-CH_2-CH-COO^-$ $\quad\quad\quad\quad\quad\quad +NH_3$	2.2	9.1	10.1
Asparagine (Asn)	$H_2N-C-CH_2-CH-COO^-$ $\quad\quad \| \quad\quad\quad +NH_3$ $\quad\quad O$	2.0	8.8	
Glutamine (Gln)	$H_2N-C-CH_2-CH_2-CH-COO^-$ $\quad\quad \| \quad\quad\quad\quad\quad +NH_3$ $\quad\quad O$	2.2	9.1	
Cysteine (Cys)	$CH_2-CH-COO^-$ $\| \quad\quad +NH_3$ SH	1.7	10.8	8.3
Amino acids with charged polar R groups				
Aspartate (Asp)	$^-OOC-CH_2-CH-COO^-$ $\quad\quad\quad\quad +NH_2$	2.1	9.8	3.9
Glutamate (Glu)	$^-OOC-CH_2-CH_2-CH-COO^-$ $\quad\quad\quad\quad\quad\quad +NH_3$	2.2	9.7	4.3
Arginine (Arg)	$H-N-CH_2-CH_2-CH_2-CH-COO^-$ $\quad\| \quad\quad\quad\quad\quad\quad\quad +NH_3$ $\quad C=NH_2^+$ $\quad\|$ $\quad NH_2$	2.2	9.0	12.5
Lysine (Lys)	$CH_2-CH_2-CH_2-CH_2-CH-COO^-$ $\|\quad\quad\quad\quad\quad\quad\quad\quad +NH_3$ $+NH_3$	2.2	9.0	10.5
Histidine (His)	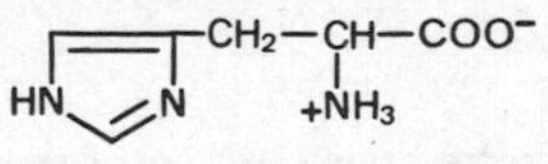 $\quad\quad\quad-CH_2-CH-COO^-$ $HN \diagdown N \quad +NH_3$	1.8	9.2	6.0

*The structure shown is that which predominates at pH 7.4.

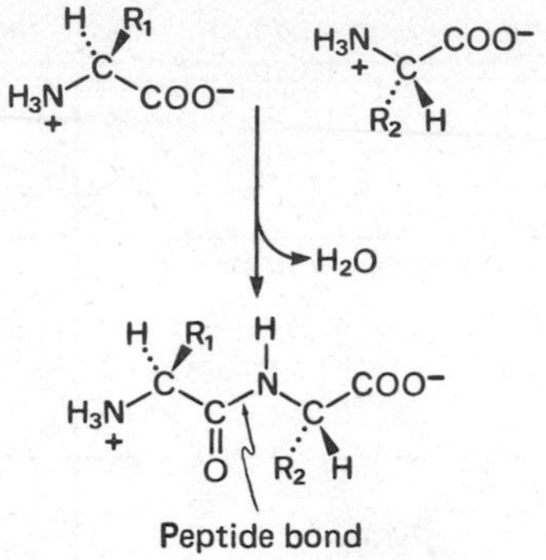

Peptide bond

Figure 3–2. Peptide bond formation.

Peptide Bonds

Proteins are unbranched polymers. During polymerization, the α-amino group of one amino acid reacts with the α-carboxyl group of another to form an amide linkage known as a peptide bond (Fig 3–2). For this reason, proteins are also referred to as **polypeptides.**

Two different resonance structures can be drawn for the peptide bond (Fig 3–3A). In reality, the structure is intermediate between these 2 extremes, and the resonance electrons are distributed over both the C–O and the C–N bonds (Fig 3–3B). Owing to resonance, the peptide bond assumes a partial double bond character, and both the carbonyl and imino groups carry partial charges. This has 2 important consequences for protein structure. (1) The

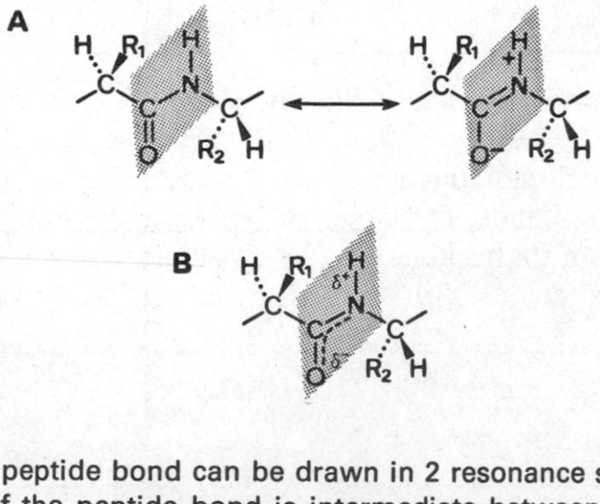

Figure 3–3. A: The peptide bond can be drawn in 2 resonance structures. **B:** The actual structure of the peptide bond is intermediate between these 2 extreme structures. Atoms that participate in the resonance structures of the peptide bond are shaded. The dashed line indicates that the peptide bond has a character intermediate between a single and a double bond. The symbol δ indicates that the charges on the atoms of the peptide group are partial.

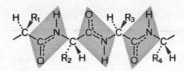

Figure 3–4. The general structure of a polypeptide. The atoms of the rigid planar peptide groups are shaded. The polypeptide chain is free to rotate only at the single bonds that connect the α-carbon atoms to the atoms of the peptide bond.

partial charges on the peptide CO and NH groups allow them to serve as H bond acceptor and H bond donor, respectively. (2) The partial double bond nature of the peptide bond prevents its free rotation, with the result that all of the atoms between one α carbon and the next lie in the same plane (Fig 3–4).

Because proteins are unbranched polymers of amino acids connected by peptide bonds, they may be represented as shown in Fig 3–4. The polypeptide chain can be viewed as having 2 components: a monotonous repeating backbone, in which α carbons and peptide bonds alternate; and variable amino acid side chains that are attached to the α carbons. All of the α-amino groups except that of the *first* amino acid and all of the α-carboxyl groups except that of the *last* amino acid are involved in peptide bonds. Thus, each protein chain has only one free α-amino group and one free α-carboxyl group, at the N and C termini of the chain, respectively. In representing polypeptides, it is customary to place the N terminus on the left.

Polypeptide Folding

Despite the restrictions on rotation imposed by the planarity of the peptide bond, polypeptide chains have a great deal of flexibility. Because the α carbon of each amino acid residue is joined to the adjacent atoms in the chain by single bonds, the chain is free to rotate around two-thirds of the bonds that form its backbone. Owing to this conformational freedom, polypeptide chains are able to fold into a variety of shapes. In its natural environment, each protein tends to assume only one conformation, designated its **native conformation.** The pattern into which a protein folds is shaped by both restrictions and attractions. Both the rigidity of the peptide bonds and steric hindrance between side chains prevent the polypeptide chain from adopting certain conformations. The conformation that is adopted is the one that **maximizes the number of noncovalent bonds** formed by both the protein and the solvent in which it is found. The explanation of how noncovalent bonds shape protein structure is described in the following paragraphs.

⬧ Proteins found in an aqueous (ie, polar) environment are termed "soluble proteins." A soluble protein generally folds into a globular shape in which most of its polar side chains are on the surface of the protein and thus in contact with water, and most of its nonpolar side chains are gathered together in a nonpolar phase that forms the core of the folded protein. Segregation of the nonpolar groups into a separate hydrophobic phase is favored thermodynamically because it permits formation of the maximum number of H bonds among the surrounding water molecules. Each water molecule is able to form H bonds to other water molecules and to the polar groups of the protein but not to the nonpolar groups. If the nonpolar side chains of a protein remain in contact with water, they limit the number of H bonds that can be formed by water molecules in their vicinity. The thermodynamic advantage that is gained by forming a hydrophobic core is the *major* force that confers conformational stability on globular proteins.

In the process of folding the nonpolar side chains into the core of the protein, portions of the polypeptide backbone are, of necessity, also buried. Because the backbone includes the polar groups of the peptide bond, the folding process removes some polar groups from the solvent. This is thermodynamically feasible *only* if each polar group placed in the core can bond to another polar group. Thus, the conformation adopted by the polypeptide chain will be determined in part by the necessity of finding partners for each of the buried polar groups.

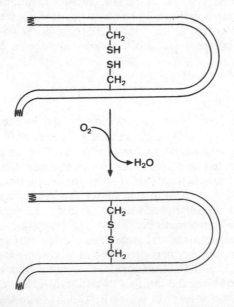

Figure 3–5. Two cysteine side chains react to form a disulfide bond. The polypeptide backbone is represented by a continuous band.

The native conformation of a protein that remains inside the cell in which it is synthesized is stabilized solely by noncovalent bonds. That of an extracellular protein may be additionally stabilized by a type of covalent cross-link, termed a disulfide bond or S–S bridge (Fig 3–5). Disulfide bonds are formed when 2 cysteine residues, brought into proximity by the folding process, are oxidized. The 2 cysteine residues that pair to form a disulfide bond may be part of the same chain or of different polypeptide chains. Disulfide bonds can thus be seen as the final locks on the 3-dimensional structure initially formed by noncovalent bonds.

Levels of Protein Structure

Proteins are described in terms of 4 levels of structure: primary, secondary, tertiary, and quaternary.

The **primary structure** of a protein is its linear sequence of amino acids from N terminus to C terminus.

The primary structure of a protein determines identity. The human body contains many thousands of species of proteins, such as hemoglobin, the carrier of oxygen in the blood, and trypsin, a digestive enzyme produced by the pancreas. Each protein species differs from all others in its primary structure. Conversely, all of the molecules of a given species of protein have the same primary structure. The primary structure also dictates the folding of the polypeptide chain. Thus, the primary structure of a protein controls its secondary, tertiary, and quaternary structures.

Secondary structure consists of regular local patterns of folding within a portion of a polypeptide chain. Secondary structures are primarily stabilized by H bonds between the NH and CO groups of the peptide bonds. A polypeptide tends to form secondary structures because of the regularity of the backbone of the chain and because the secondary structures maximize the number of H bonds that can be formed. In a typical protein, more than 60% of the amino acid residues are involved in 3 types of secondary structure—helices, pleated sheets, and reverse turns.

If at every α carbon the polypeptide chain twists by the same amount, the backbone will follow a helical path. The most common type of helix found in polypeptides, the α **helix** (Fig 3–6), consists of a regular right-handed helix in which the NH group of each peptide bond is bonded to the CO group of the residue located 4 positions later in the chain. In an α helix, all of the peptide bond CO and NH groups are involved in H bonds lying *parallel* to the axis of the helix. The polypeptide backbone forms the core of the helix, while the amino acid side chains project to the outside and coat its surface. The chemical features of the surface of the helix thus depend on the identities and order of the amino acids in the primary structure. Because the peptide N of proline is a member of a ring and thus lacks a substituent H for H bonding, proline residues terminate helices. The α helix

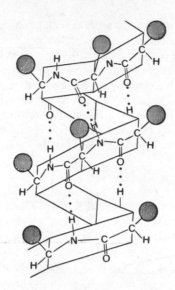

Figure 3–6. Model of an α helix. To emphasize the helical path taken by the polypeptide chain, the backbone has been superimposed on a ribbon. Intrachain H bonds are represented by dotted lines. The amino acid side chains are represented by solid spheres. (Redrawn, with permission, from Barker R: *Organic Chemistry of Biological Compounds.* Prentice-Hall, 1971.)

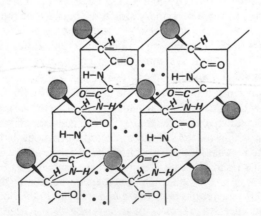

Figure 3–7. A parallel β-pleated sheet formed by 2 adjacent polypeptide chains. To emphasize the path taken by the polypeptide chain, the backbone has been superimposed on a ribbon. (Redrawn, with permission, from Barker R: *Organic Chemistry of Biological Compounds.* Prentice-Hall, 1971.)

is also terminated by amino acids with large side chains and by charge repulsion between side chains of like charge.

Another common secondary structure is the β-pleated sheet (Fig 3–7). In this conformation, the polypeptide chain is almost fully extended, and successive amino acid side chains project outward on opposite sides of the backbone. The β-pleated sheet is favored when 2 or more sections of chain can be aligned next to each other on the same axis so as to form interstrand H bonds. As the name suggests, the sheet is not flat but has a slightly pleated surface. The H bonds that hold the sheet together are *perpendicular* to the polypeptide backbone, and the side chains project above and below the surface of the sheet. The N to C orientation of adjacent strands may be the same (parallel) or opposite (antiparallel).

The backbone of a polypeptide chain bends back on itself at frequent intervals. Abrupt turns are made possible both by proline residues and by β turns, secondary structures in which the CO group of one amino acid residue forms an H bond with the NH group of the residue 3 positions later in the chain (Fig 3–8). Formation of a β turn is favored at those positions in the polypeptide chain that contain several polar amino acids in succession.

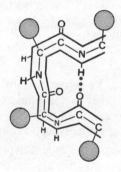

Figure 3–8. A β turn allows the polypeptide chain to bend back on itself.

Tertiary structure is the overall folding of a single polypeptide chain. Fig 3–9 presents 2 models of a globular protein called bovine pancreatic ribonuclease. The space-filling model is useful for representing the shape of the protein and emphasizes the fact that globular proteins are very compact. The α-carbon model, in which only the α-carbon atoms of the polypeptide backbone are represented, reveals the internal structure of the protein but gives the incorrect impression that there are open spaces within the protein structure. In fact, there is almost no space within the center of a globular protein.

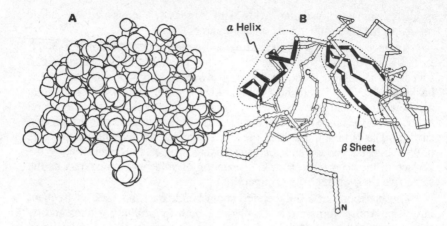

Figure 3-9. Two models of bovine pancreatic ribonuclease viewed from the same orientation. **A:** A space-filling model. Each sphere represents an atom in the structure of ribonuclease. H atoms are not depicted. **B:** An α-carbon model. This model shows only the course of the polypeptide backbone. Each α carbon of the polypeptide chain is represented by a circle. Rods represent the atoms and bonds that span the distance from one α carbon to the next. Disulfide bonds are represented as −S−S−. Dotted lines enclose an area of α helix and one of β-pleated sheet. (Computer-generated models courtesy of S Sprang.)

As the protein folds into its native conformation, amino acid residues that are distant from each other in the primary structure are juxtaposed. Within the tertiary structure are areas of both α and β structures linked by irregularly folded portions of the polypeptide chain.

In the case of ribonuclease, a single polypeptide chain constitutes the complete protein. Many other proteins are made up of more than one polypeptide chain. **Quarternary structure** is the arrangement of polypeptides together in a multichain complex. Complexes of polypeptides are held together by the same types of bonds that determine the folding of an individual polypeptide. The surfaces along which 2 polypeptides interact are fitted to each other, showing complementarity of shape, charge, and polarity.

Assembling a protein from multiple polypeptide chains offers biologic advantages in the form of economy and control. Viruses have employed this strategy to construct protective coats made up of multiples of one or a few types of polypeptide. In doing so, the virus uses a small amount of genetic information to specify a large structure made up of repeating subunits. As will be seen in the next chapter, polypeptide complexes also increase the sensitivity with which the activity of a protein can be controlled.

Denaturation

The biologic activity of a protein depends on the proper 3-dimensional arrangement of its functional groups. Thus, proteins are able to function only when folded in their native conformations. Because the native conformations of most proteins are stabilized solely by noncovalent bonds, folding may be disrupted, or denatured, by conditions such as high temperature, extremes of pH, or the presence of organic solvents that weaken noncovalent interactions. Denaturation disrupts the biologic activity of a protein as well as its secondary, tertiary, and quaternary structures but leaves the primary structure intact. The formation of disulfide bonds in a protein increases its resistance to denaturation.

4 | Protein Function: Binding

OBJECTIVES

- Be able to explain the mechanism by which hemoglobin and myoglobin reversibly bind oxygen.

- Be able to explain how the structures of myoglobin and hemoglobin limit the affinity of heme for carbon monoxide.

- Be able to interpret the oxygen-binding curves for myoglobin and hemoglobin.

- Know the definitions of cooperativity and allostery. Be able to explain how the quaternary structure of hemoglobin accounts for its cooperativity in binding oxygen.

- Know what the Bohr effect is and what causes it. Be able to explain the physiologic utility of the Bohr effect.

- Know how bisphosphoglycerate affects hemoglobin's affinity for oxygen and how this contributes to adaptation to high altitude.

- Know how hemoglobin S differs from hemoglobin A and how this change accounts for anemia in individuals homozygous for the hemoglobin S gene.

MANY OF THE proteins of the body bind other molecules and thereby serve as molecular carriers. This chapter describes how 2 related proteins—hemoglobin and myoglobin—bind oxygen and thus mediate oxygen storage and transport.

Because the solubility of O_2 in water is low and the rate at which it enters tissues by simple diffusion is slow, large aerobic organisms could not have evolved without a circulatory system and a carrier of oxygen. In vertebrates, the carrier within the blood system is hemoglobin, the major protein of red blood cells. A related protein, myoglobin, functions in red muscle to store O_2 absorbed from the blood and release it as needed to the intracellular site of oxidation, the mitochondrion. Both hemoglobin and myoglobin bind oxygen, but they operate at different oxygen partial pressures (P_{O_2}) and in different gradients of oxygen tension. Hemoglobin takes up oxygen in the capillaries of the lung, where the P_{O_2} is 100 mm Hg, and discharges it in the capillaries of the peripheral tissues at a P_{O_2} of 20 mm Hg. Myoglobin operates in a steeper gradient of oxygen tensions, binding oxygen at the tissues' capillary oxygen partial pressure and discharging it to the mitochondrion when the oxygen partial pressure there is low.

Structure of Myoglobin

Myoglobin has 2 components, both of which are essential to its role as an oxygen carrier: (1) globin, a polypeptide; and (2) heme, a planar ring compound that has ferrous iron (Fe^{2+}) at its center (Fig 4–1). Heme is a member of the class of compounds called porphyrins (see Chapter 11). The role of heme in myoglobin is to bind O_2; that of globin is to make O_2 binding reversible. In myoglobin and all other heme-containing proteins, heme func-

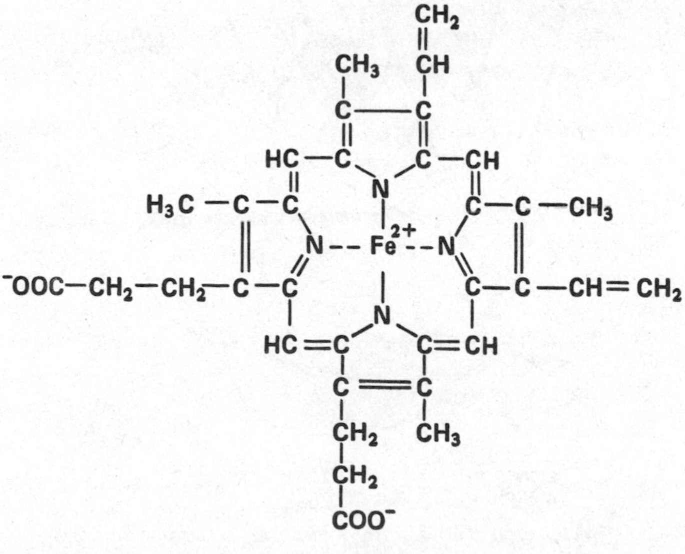

Figure 4–1. The structure of heme.

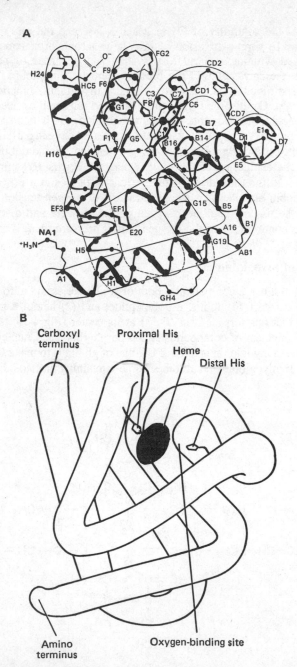

Figure 4–2. A: An α-carbon model of myoglobin. **B:** A tubular model of myoglobin. (**A** is based on Dickerson RE in: *The Proteins,* 2nd ed. Vol 2. Neurath H [editor]. Academic Press, 1964. Reproduced with permission.)

tions as a **prosthetic group,** ie, a nonpolypeptide part of the protein that participates directly in the function of the protein.

Two models of myoglobin are shown in Fig 4–2. The predominant secondary structure feature of the single polypeptide chain of myoglobin is the α helix. Folding of the globin polypeptide produces 8 helical regions (designated A through H, starting at the N terminus), which in turn come together to form a pocket that holds the heme prosthetic group. The heme-binding pocket is formed by amino acid residues which, although distant from each other in the primary structure, are brought together by the folding process. The heme group, which is largely nonpolar, is held in place in the globin pocket by contacts with the side chains of a number of nonpolar amino acids, including alanine, valine, leucine, isoleucine, and phenylalanine. Globin *specifically* binds heme (and does not bind other small molecules) because the pocket is the right shape for the heme and because the amino acid side chains in contact with the ring are nonpolar. **As a general rule, a binding site on a protein fits the compound to be bound in size, shape, and polarity.**

Binding of Oxygen & Carbon Monoxide

Fe^{2+} is able to form bonds to 6 other atoms. In heme, 4 of these bonds are made to the nitrogen atoms of the porphyrin ring and are in the plane of the ring (Fig 4–3). The remaining 2 project out from either side of the

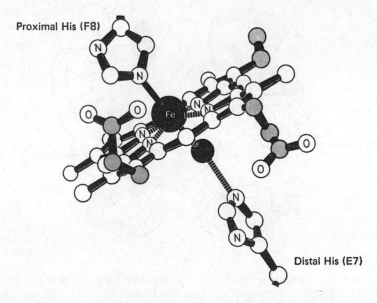

Proximal His (F8)

Distal His (E7)

Figure 4–3. Oxygen binds to the heme prosthetic group of myoglobin. (Reproduced, with permission, from Löffler G et al: *Physiologische Chemie.* Springer-Verlag, 1979.)

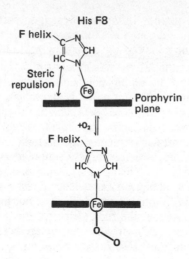

Figure 4–4. Oxygen binding changes the shape of myoglobin. (Slightly modified and reproduced, with permission, from Stryer L: *Biochemistry,* 2nd ed. Freeman, 1981.)

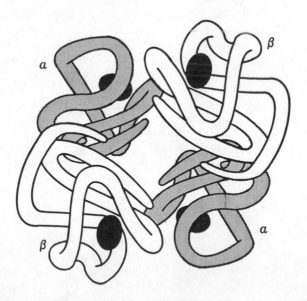

Figure 4–5. The quaternary structure of hemoglobin. Each of the 4 polypeptide chains of hemoglobin is represented by a continuous band. The heme groups are represented by black disks. (Redrawn, with permission, after Dickerson RE: X-ray studies of protein mechanisms. *Annu Rev Biochem* 1972;**41**:815. Copyright © 1972 by Annual Reviews Inc.)

ring structure. When heme is part of the myoglobin structure, one bond is made to a histidine side chain (His F8—so named because it is the eighth residue in helix F). The sixth bond is made to oxygen. The oxygen molecule also forms a bond to another histidine side chain (His E7). His F8 and His E7 are also called the proximal and distal histidines, respectively.

An isolated heme molecule can combine with oxygen, but when it does so the iron atom is irreversibly oxidized from the Fe^{2+} to the Fe^{3+} state. An intermediate in this process is a compound in which O_2 is sandwiched between 2 heme moieties. One of the functions of the globin polypeptide chain is to prevent heme units from coming close enough to each other to form this intermediate. For heme to function as a reversible carrier of oxygen, the interaction between the heme iron and O_2 must also be made weaker. Myoglobin weakens the attraction of Fe^{2+} for O_2 by means of the interaction between His F8 and the iron atom. By donating electrons to the iron atom, the histidine residue makes it more electronegative and thereby weakens its attraction for oxygen.

Carbon monoxide (CO) competes with oxygen for binding to the heme iron and thus is toxic because it limits the transport and storage of oxygen. A major role of the distal histidine residue (His E7) of globin is to reduce the affinity of heme for CO. By itself, heme binds CO 25,000 times more tightly than it does O_2. But when heme is complexed with the globin protein, the distal histidine residue sterically hinders optimal binding of CO, although it does not interfere with O_2 binding.

Oxygen binding alters the myoglobin tertiary structure. When the O_2-binding site of myoglobin is empty, the iron atom of heme lies slightly outside of the porphyrin ring in the direction of His F8 and the histidine side chain is slightly off center in its relationship to the ring (Fig 4–4). This is the most stable conformation of the unoccupied protein. When oxygen binding occurs, the iron atom is pulled into the plane of the porphyrin ring, drawing the histidine side chain and the α helix of which it is a part (helix F) into a new position. The energy gained from binding oxygen is sufficient to counter the strain placed on the protein conformation by this movement. Myoglobin is not an exceptional case; **many proteins undergo a small conformational change upon binding another molecule.**

Structure of Hemoglobin

Humans produce several types of hemoglobin. Although they differ somewhat in their primary structures, they all share the same general architecture and mode of action. All of the human hemoglobins are multimeric proteins that contain 4 polypeptide chains, 2 each of 2 types. The polypeptides of the major adult human hemoglobin, hemoglobin A, are designated α and β. The arrangement of the chains in the tetramer is as shown in Fig 4–5.

The tertiary structures of the α and β polypeptides are similar to that of myoglobin. Each hemoglobin polypeptide has 8 helical regions that fold

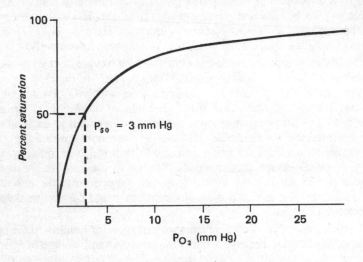

Figure 4-6. The oxygen saturation curve of myoglobin.

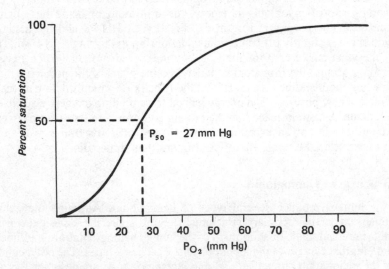

Figure 4-7. The oxygen saturation curve of hemoglobin.

to form a nonpolar heme pocket. Thus, each subunit has an oxygen-binding site. A major difference between the myoglobin chain on the one hand and the hemoglobin chains on the other lies in the surfaces of the tertiary structures. Myoglobin's surface consists largely of polar residues that confer water solubility upon the protein. The hemoglobin polypeptides have surfaces that are complementary to each other and lead to their aggregation first as an $\alpha\beta$ dimer and then as an $(\alpha\beta)_2$ tetramer.

Oxygen Saturation Curves

In order to understand the functional differences between myoglobin and hemoglobin, it is useful to compare their oxygen saturation curves (Figs 4–6 and 4–7). Two differences in the curves are immediately apparent. (1) The oxygen tension required for 50% saturation of the myoglobin (P_{50}) is lower than that required for the same degree of saturation of hemoglobin; ie, myoglobin has a greater affinity for oxygen than does hemoglobin. This difference permits myoglobin to bind oxygen as it is released from hemoglobin and thereby serve as a reservoir for oxygen in red muscle. (2) The 2 curves also differ in shape. The oxygen saturation curve of myoglobin is a rectangular hyperbola, while that of hemoglobin is sigmoid. Myoglobin's hyperbolic curve indicates that its affinity for oxygen is independent of oxygen concentration. The sigmoid curve of hemoglobin indicates that its affinity for oxygen increases as the oxygen concentration increases. The significance of this difference becomes apparent when we ask how large a change in oxygen tension is required to fully unload oxygen from each of these proteins. Fig 4–6 demonstrates that at the P_{O_2} of arterial blood in resting muscle, about 20 mm Hg, approximately 85% of myoglobin is loaded with oxygen. In order to unload most of the bound oxygen, the oxygen concentration in the environment of myoglobin must decrease more than 100-fold. In other words, the full oxygen-binding capacity of myoglobin can be utilized only if there is a substantial change in oxygen tension. Hemoglobin encounters oxygen tensions that vary, at most, only 5-fold. If the properties of hemoglobin resembled those of myoglobin, only a very small part of its oxygen-binding capacity would be used. However, the affinity of hemoglobin for oxygen varies 300-fold in the range of oxygen concentrations found in circulation, and thus a much larger part of the oxygen bound in the lungs can be released in the peripheral tissues.

The sigmoid shape of hemoglobin's oxygen saturation curve is a consequence of its multimeric structure. When one subunit of hemoglobin binds oxygen, the conformation of the protein is changed in a way that increases the affinity of the remaining 3 subunits for oxygen. This phenomenon, termed **positive cooperativity,** depends on the integrity of the hemoglobin quaternary structure. If the tetramer is disaggregated into independent monomeric subunits, cooperativity is abolished and the oxygen-binding curve becomes hyperbolic in shape.

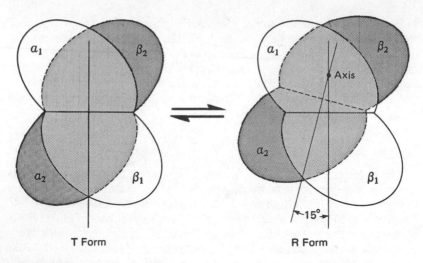

Figure 4–8. Hemoglobin has 2 quaternary structures, called T and R. (Reproduced, with permission, from Martin DW Jr et al: *Harper's Review of Biochemistry*, 20th ed. Lange, 1985.)

Mechanism of Cooperativity

The positive cooperativity demonstrated by hemoglobin is mediated by changes in its quaternary structure. X-ray crystallographic data show that hemoglobin has 2 quaternary structures, ie, 2 ways in which the subunits can pack together (Fig 4–8). The 2 forms, designated the T (taut), or deoxy, form and the R (relaxed), or oxy, form, differ in the type and strength of bonds linking the subunits. The quaternary structure of the T form is stabilized by a larger number of ionic bonds than is that of the R form. For this reason, in the absence of other factors, T predominates.

The T and R forms also differ in their affinities for oxygen. While R binds O_2 as avidly as does myoglobin, the oxygen affinity of T is approximately 300-fold lower. The difference in affinities is a direct consequence of the different quaternary structures. Hemoglobin, like myoglobin, changes shape upon binding oxygen. As the heme group binds O_2, the Fe^{2+} ion moves into the plane of the porphyrin ring and the F helix is shifted. The ionic bonds that stabilize the T form impede the movement of the F helix and the Fe^{2+} ion. Thus, oxygen binds less easily to the T form than it does to either the R form or myoglobin. However, when an O_2 molecule does bind to T and movement of the F helix and Fe^{2+} takes place, the ionic bonds that stabilize the T structure are strained. Each O_2 that binds adds to that strain, favoring a transition to the R form, which has a higher affinity for oxygen (Fig 4–9). Conversely, when one or more oxygen molecules dissociates from hemoglobin, the strain favoring the R form is lessened and the tetramer tends to switch back to the T form.

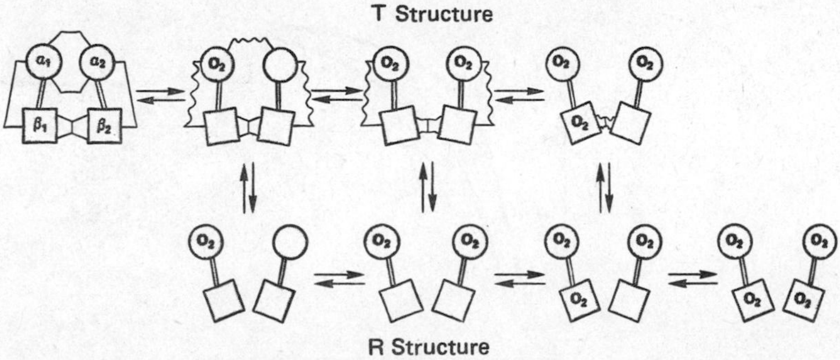

T Structure

R Structure

Figure 4–9. Oxygen binding favors the transition from T to R. (Modified and redrawn, with permission, from Perutz MF: Hemoglobin structure and respiratory transport. *Sci Am* [Dec] 1978;**239**:92.

We can summarize this model with 3 statements: (1) Hemoglobin has 2 quaternary structures that differ in their O_2 affinities. (2) In the absence of oxygen, the low-affinity form is more stable. (3) Oxygen binding shifts the equilibrium in favor of the high-affinity form.

Allosteric Effectors of Oxygen Binding

CO_2 and H^+ are metabolic by-products and therefore are present in relatively high concentrations in active peripheral tissues, such as contracting muscle. Hemoglobin facilitates their removal by binding them in the peripheral tissues and releasing them in the lungs. In binding H^+, hemoglobin provides a buffer against changes in blood pH. Furthermore, H^+ and CO_2 decrease the affinity of hemoglobin for oxygen; ie, their presence shifts the oxygen binding curve to the right. Thus, when oxygenated hemoglobin encounters a low-pH, high-CO_2 environment, it tends to release its bound oxygen. This serves to increase the delivery of O_2 to metabolically active tissues. The decrease in hemoglobin's oxygen affinity with a decrease in pH is called the **Bohr effect** (Fig 4–10).

The effect of pH on oxygen binding is mediated by changes in hemoglobin's quaternary structure. Protons influence the quaternary structure of the tetramer because they play a role in the formation of the intersubunit ionic bonds. Several of the ionic bonds that stabilize the T form of hemoglobin include groups, such as histidine side chains, that have pK values within the range of pH encountered in circulation. In the low-pH environment of the peripheral tissues, these groups become protonated. This in turn leads to the formation of the T ionic bonds and facilitates the dissociation of oxygen from the protein. At the higher pH of the lungs, protons are released and the ionic bonds are broken, thus promoting a transition to the R form and binding of oxygen.

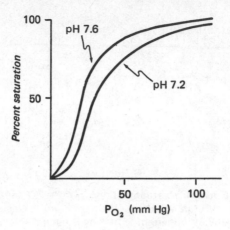

Figure 4–10. The Bohr effect.

The interaction of CO_2 with hemoglobin also affects the T to R transition. CO_2 reversibly carbamoylates the amino termini of the hemoglobin polypeptide chains.

$$Hb—NH_2 + CO_2 \rightleftharpoons Hb—NH—COO^- + H^+$$

When carbamoylated, the termini are negatively charged and are able to form ionic bonds that help stabilize the T quaternary structure.

Bisphosphoglycerate (DPG; formerly diphosphoglycerate) (Fig 4–11), a phosphorylated carbohydrate made in the red blood cell, is part of the physiologic mechanism for adapting to changes in oxygen availability. When delivery of oxygen is impaired, eg, by high altitude or lung disease, the red blood cell increases its production of DPG. DPG binds only to the T form of hemoglobin, fitting into a positively charged cavity between the β chains (Fig 4–12). DPG binding stabilizes the T form of hemoglobin and thus decreases its oxygen affinity.

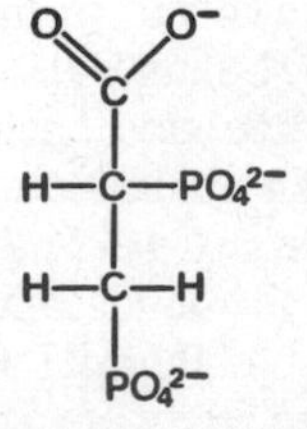

Figure 4–11. The structure of DPG.

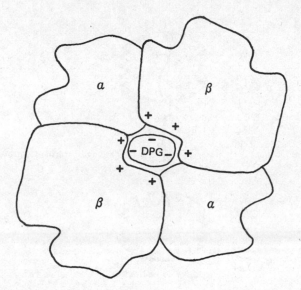

Figure 4-12. DPG fits within a positively charged cavity between the β subunits of hemoglobin.

How could a decrease in oxygen affinity be an adaptive response to insufficiency of oxygen? The answer lies in the observation that hemoglobin's oxygen affinity is normally somewhat too high to achieve maximal oxygen release in peripheral tissues. As DPG shifts the oxygen saturation curve to the right, oxygen loading in the lungs decreases slightly but oxygen unloading in the peripheral tissues increases to more than compensate. Thus, an increase in the DPG content of the red blood cell causes hemoglobin to deliver a larger portion of its bound oxygen to the tissues.

Table 4-1 summarizes the effects of O_2, H^+, CO_2, and DPG on hemoglobin. H^+, CO_2, and DPG all change the affinity of hemoglobin for oxygen by binding to sites other than the O_2-binding site. This phenomenon is called **allostery** (Greek *allos* = other), and H^+, CO_2, and DPG are termed **allosteric effectors** of O_2 binding. Allostery, like cooperativity, is mediated by changes

Table 4-1. The effects of ligand binding on the affinity of hemoglobin for O_2.

Ligand	Number of Sites per Tetramer	Quaternary Structure Stabilized	Effect of Ligand on O_2 Affinity
O_2	4	R	Increase
H^+	~2	T	Decrease
CO_2	4	T	Decrease
DPG	1	T	Decrease

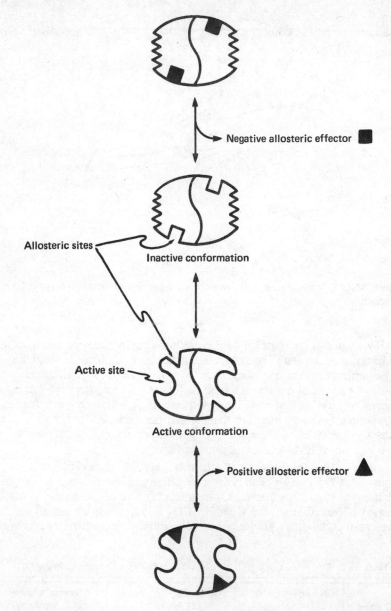

Figure 4–13. Allosteric effectors stabilize alternative conformations of an allosteric protein.

in protein conformation. All allosteric proteins have 2 or more stable conformations in equilibrium with each other. Binding of an allosteric effector stabilizes one of the conformations and thus shifts the equilibrium in favor of that conformation. Two ligands that bind to the same protein reciprocally affect each other's binding. If both bind to the same conformation, they will favor each other's binding. If they bind to different conformations, they will discourage each other's binding (Fig 4–13).

Mutant Hemoglobins

While many changes in the amino acid sequence have no detectable effect on hemoglobin's function, others interfere with its transport of oxygen in ways that support our model of hemoglobin action. Here we will discuss only those hemoglobin mutations that result in single amino acid changes. Many of the mutants in this category can be assigned to one of 4 classes.

(1) Hemoglobin M: In order to transport oxygen, the hemoglobin iron atom must be in the divalent state. When it is part of the normal hemoglobin A molecule, Fe^{2+} is oxidized to Fe^{3+} at a slow rate. The resulting inactive hemoglobin is called **methemoglobin.** Five mutant hemoglobins (called hemoglobin M) have structures that favor more rapid oxidation of the iron atom. In each of these mutants, one of the amino acids that makes contact with the heme iron has been replaced by another amino acid that stabilizes the iron atom in the oxidized form.

(2) Mutant hemoglobins characterized by a higher than normal affinity for oxygen: Most of these mutations change amino acids that form the intersubunit contacts. The replacement amino acids disrupt the contacts and prevent communication between the subunits. As might be expected, these mutant proteins do not display cooperativity in oxygen binding or a Bohr effect.

(3) Hemoglobin S (sickle hemoglobin): Replacement of a single amino acid of the β chain (Glu A2 → Val) results in a hemoglobin with altered assembly properties. In the deoxy state, hemoglobin S aggregates into long filaments that distort the red blood cell into a characteristic sickle shape. When sickled, the cells are unable to pass through capillaries, and delivery of oxygen to peripheral tissues is thus impaired. Sickled cells are also more rapidly removed from circulation by the spleen than are cells of the normal shape. Individuals who are homozygous (genotype S/S) for the hemoglobin S gene therefore have fewer red blood cells in circulation and are anemic. Individuals who are heterozygous for hemoglobin S (genotype A/S) carry the trait but generally are not anemic. The gene for hemoglobin S has a high incidence in populations historically exposed to malaria. In the heterozygote, the hemoglobin S gene product provides some protection against the malaria parasite that develops within the red blood cell. Because infected cells require larger amounts of oxygen than uninfected ones, they tend to become sickled sooner and thus be removed from circulation.

(4) **Hemoglobins with decreased stability:** Many mutant hemoglobins are less stable than the normal protein and are degraded more rapidly. In most of these mutants, an amino acid replacement within the heme pocket has reduced the affinity with which the protein binds heme. The existence of these mutants suggests that the presence of the heme within the pocket stabilizes the protein conformation and prevents its destruction.

Protein Function: Catalysi

OBJECTIVES

- Be able to explain how a catalyst affects the thermodynamic parameters of a chemical reaction.

- Know what is meant by the active site of an enzyme and by reaction and substrate specificity.

- Be able to explain the differences between the "lock and key" model and the induced fit model of enzyme specificity.

- Be able to explain the mechanisms by which enzymes catalyze reactions.

- Be able to describe the role of a coenzyme in an enzyme-catalyzed reaction.

- Be able to explain the mechanisms by which the activity of an enzyme can be regulated.

IN VIVO, MOST reactions are catalyzed by a class of proteins called **enzymes.** In order to understand how proteins function as catalysts, it is useful to briefly review the events that occur during a chemical reaction. The following explanation of how a reaction proceeds is known as the transition state model.

During a chemical reaction, the reacting molecules collide and enter into a transition state, ie, a combination of reacting molecules intermediate between reactants and products. The transition state is short-lived and rapidly breaks down to either products or reactants. If we could measure the free energy (G) content of the transition state, we would find that it is greater

...lan that of the average reactants or the average products (Fig 5– 1). For the reaction to take place, the reacting molecules must have enough energy to enter into the transition state. The difference in free energy of the reactants and the transition state is termed the **free energy of activation** ($\Delta G^{\ddagger}$).

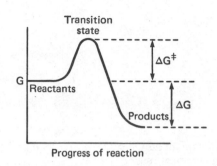

Figure 5–1. Free energy changes during the progress of a reaction.

The rate at which any reaction occurs is determined by the magnitude of the $\Delta G^{\ddagger}$ of the reaction and the energy of the reactants. At the temperatures of living organisms, few molecules have sufficient energy to reach the transition state, and thus most collisions between reactants are nonproductive. Most biologic reactions would therefore occur only very slowly in the absence of a **catalyst**—a substance that increases the rate at which a reaction occurs without itself being changed. Catalysts work by decreasing the energy barrier ($\Delta G^{\ddagger}$) between reactants and products, thereby making it easier to reach a transition state. Catalysts do *not* change the free energy difference (ΔG) between reactants and products and therefore do not affect the outcome of the reaction. A catalyst accelerates both the forward and reverse reactions and thus only increases the rate at which a reaction approaches equilibrium.

Enzyme Specificity

Enzymes are highly specific both in the reactions they catalyze and in the compounds (termed **substrates**) on which they act. We can illustrate these types of specificity by reference to 3 enzymes that degrade proteins: trypsin, chymotrypsin, and elastase. Each of these enzymes catalyzes only one type of reaction, the hydrolysis of peptide bonds (Fig 5–2). All 3 are also selective with respect to the substrates on which they act (Fig 5–3). Trypsin cleaves only those peptide bonds on the carboxyl side of arginine and lysine residues. Chymotrypsin is specific for large hydrophobic or aromatic amino acids (phenylalanine, tyrosine, and tryptophan), whereas elastase attacks only peptide bonds adjacent to small amino acids (glycine, alanine, and serine). Most enzymes also display optical specificity and act

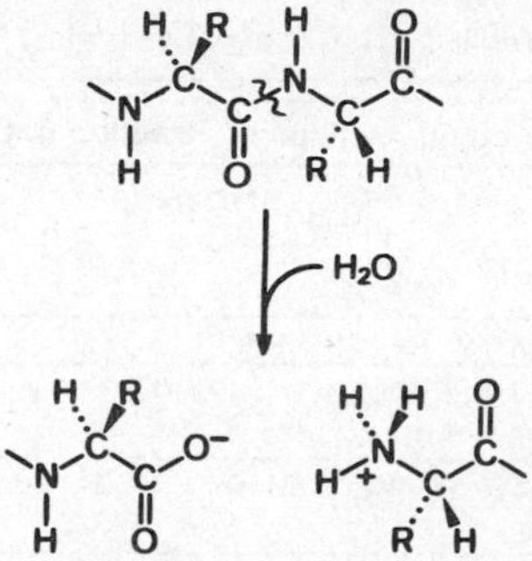
H R H O
N C C N C C
H O R H
H_2O
H R
N C C O^-
H O
H H O
H N C C
R H

Figure 5–2. Hydrolysis of a peptide bond.

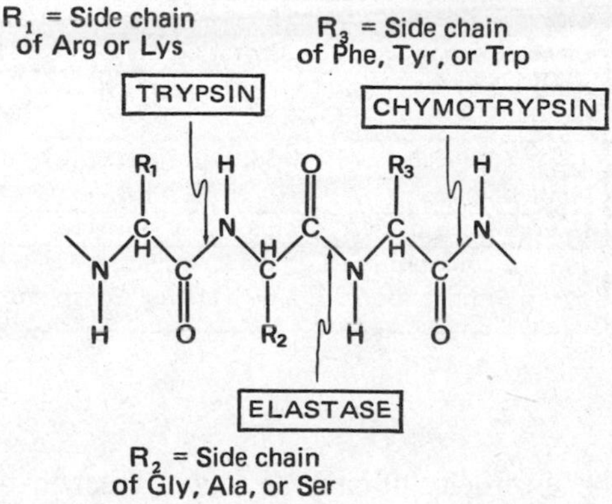
R_1 = Side chain of Arg or Lys
R_3 = Side chain of Phe, Tyr, or Trp
TRYPSIN
CHYMOTRYPSIN
ELASTASE
R_2 = Side chain of Gly, Ala, or Ser

Figure 5–3. The substrate specificities of trypsin, chymotrypsin, and elastase.

only on one isomer of a pair; trypsin, chymotrypsin, and elastase cleave polymers of L-amino acids but have no effect on polymers of D-amino acids.

Enzyme Nomenclature

The official name of an enzyme, constructed according to rules of the International Union of Biochemistry (IUB), has 2 parts. The first part names the substrates or the products of the reaction; the second designates the type of reaction catalyzed. For example, the enzyme that catalyzes an oxidation-reduction reaction using alcohol and NAD (nicotinamide adenine dinucleotide) as substrates is officially designated alcohol dehydrogenase. Table 5–1 lists the classes of enzymatic reactions that serve as the basis for the IUB nomenclature.

Table 5–1. Classes of enzymes.

Class of Enzyme	Type of Reaction Catalyzed
Oxidoreductase, dehydrogenase, or reductase	Transfers H from one substrate to another.
Transferase	Transfers a group other than H.
Kinase	Transfers a phosphoryl group from a high-energy compound to another compound.
Mutase	Transfers a group from one position to another within the same molecule.
Phosphorylase	Cleaves a bond by adding inorganic phosphate (P_i) to it.
Hydrolase	Cleaves a bond by the addition of water.
Phosphatase	Hydrolyzes substrates to liberate P_i.
Lyase	Removes groups, by mechanisms other than hydrolysis, leaving double bonds.
Aldolase	Cleaves a double bond to form an aldehyde group.
Decarboxylase	Cleaves carboxyl groups, liberating them as CO_2.
Hydratase	Adds water to a double bond or removes water, forming a double bond.
Synthase	Joins 2 molecules without using the energy of a high-energy phosphate compound.
Isomerase	Interconverts optical, geometric, or positional isomers.
Ligase or synthetase	Joins 2 compounds using the energy released in the hydrolysis of a pyrophosphate bond of a high-energy phosphate compound.

Active Sites

To explain the specificity of enzyme action, Fischer in 1890 proposed that the substrates of an enzymatic reaction bind to a specific site on the enzyme and that the shape of that site is complementary to that of the substrates (Fig 5–4). His hypothesis is known as the "lock and key" model of enzyme specificity. Modern biochemistry has largely substantiated this view.

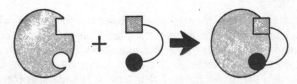

Figure 5–4. The "lock and key" model of enzyme specificity. (Reproduced, with permission, from Martin DW Jr et al: *Harper's Review of Biochemistry*, 20th ed. Lange, 1985.)

That part of an enzyme to which the substrates bind and at which catalysis takes place is called the active site. In most cases, the active site is a pocket or groove in the surface of the protein. Included within the pocket are functional groups that attract the substrate molecules and mediate the catalytic event. Like the heme-binding pocket of myoglobin, the active site is made up of amino acid residues that may be distant from each other in the primary structure of the protein but are brought together as the polypeptide chain folds into its tertiary structure.

X-ray crystallographic studies have provided support for the "lock and key" model of specificity by showing that many active sites are complementary to their substrates in shape and chemical properties. The active site of trypsin, for example, includes a hydrophobic cavity that has a negative charge at its innermost point (Fig 5–5). This cavity is large enough to bind an arginine or lysine side chain, and the negative charge at the end of the cavity can form an ionic bond with the positively charged end of the side chain. Chymotrypsin has a large hydrophobic pocket that accommodates hydrophobic side chains. Elastase has almost no side chain pocket and thus excludes all but the smallest amino acids.

While these examples support a "lock and key" model, it is now clear that many active sites are not rigid but change shape upon binding their substrates. The change in shape that is caused by substrate binding (termed **induced fit**) both improves the fit of the active site to the substrate and brings the catalytic groups into the correct position for action. Induced fit accounts, in part, for the substrate specificity of trypsin. Although peptide bonds adjacent to glycine residues easily fit into the active site, these bonds are not normally substrates for trypsin. However, when ethylamine is included in the reaction, peptide bonds adjacent to glycine residues are cleaved (Fig 5–6). In the latter case, ethylamine fills the side chain binding pocket of trypsin and mimics the side chain of the normal substrate amino acids. The effect of ethylamine on trypsin suggests that occupancy of the side chain pocket by a lysine or arginine side chain or by something that resembles these groups is necessary to properly orient the catalytic groups within the active site.

Mechanisms of Catalysis

For a reaction to occur, the reactants must have sufficient energy to reach the transition state; ie, they must overcome the energy barrier posed by the $\Delta G^{\ddagger}$. A large part of the catalytic power of an enzyme depends on its ability to **lower the activation barrier** separating reactants from products. To do so, an enzyme may provide an environment within the active site that favors the transition state, or it may provide catalytic groups that allow the reaction to proceed via intermediates not part of the uncatalyzed reaction. Many enzymes act as general acid-base catalysts. In these cases, catalysis is performed by groups located within the active site that donate protons to

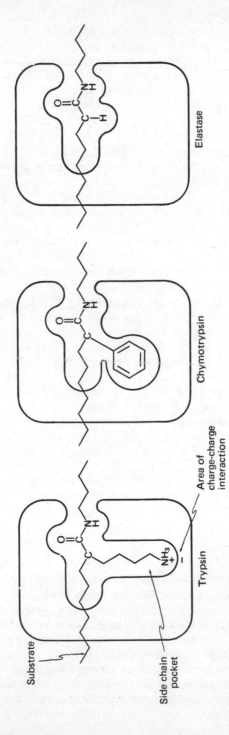

Figure 5–5. Schematic representation of the active sites of trypsin, chymotrypsin, and elastase. Each active site is complementary to its substrate in shape and properties.

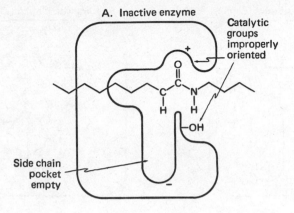

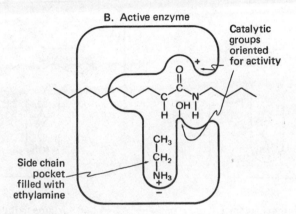

Figure 5–6. The active conformation of trypsin is formed when the side chain pocket is filled.

and accept protons from the substrate. Other enzymes operate by covalently combining with the substrate. Trypsin employs the latter mechanism to catalyze peptide bond hydrolysis. Trypsin weakens the attraction between the C and N atoms of the peptide bond by forming a reaction intermediate in which the enzyme is covalently joined to the substrate. It also uses noncovalent bonds between the active site and the transition intermediate to stabilize (ie, lower the free energy of) the intermediate.

Trypsin, chymotrypsin, and elastase belong to a family of enzymes known as the serine proteases. The enzymes of this family share a common mechanism of action that relies upon a reactive serine residue located within

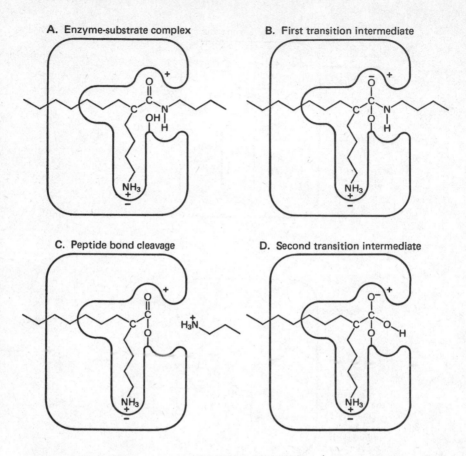

A. Enzyme-substrate complex

B. First transition intermediate

C. Peptide bond cleavage

D. Second transition intermediate

Figure 5–7. The mechanism of action of trypsin. See text for a description of the events in the hydrolysis of a peptide bond.

the active site of each enzyme (Fig 5–7). In each case, hydrolysis of the peptide bond is initiated when the serine residue forms a bond to the carbonyl carbon of the substrate peptide bond. This leads to the formation of a transition intermediate in which the substrate carbonyl carbon atom is bonded to 4 atoms and the carbonyl oxygen has an unsatisfied valency and is thus negatively charged. By itself, such a transition intermediate would be hard to achieve. However, the environment of the active site makes it more likely for this intermediate to form by providing groups that can bond with the charged portions of the transition intermediate. Although the transition intermediate is stabilized somewhat by the active site, it is nevertheless unstable and will break down. Decay of the transition intermediate results in cleavage of the peptide bond. That portion of the substrate that includes the newly formed amino group then leaves the active site. Subsequently, water enters the active site and, through a similar transition intermediate, reacts with the carbonyl carbon and detaches it from the active site of the enzyme.

In the case of the serine proteases, the catalytic groups are themselves part of the polypeptide chain. However, polypeptides cannot by themselves catalyze all of the biologically important reactions. Many enzymes depend for their activity on small molecules or metal ions that serve as **coenzymes.** A coenzyme interacts with the active site of the enzyme and acts as a catalytic group in the reaction. Some coenzymes are permanently attached, either covalently or noncovalently, to their respective enzymes. In these cases, the polypeptide-coenzyme complex is termed a **holoenzyme** (Greek *holos* = whole) and the polypeptide by itself, which is catalytically inactive, an **apoenzyme** (Greek *apo* = from or away from). Other coenzymes are only transiently bound to their enzymes and behave like substrates of the reaction.

Each type of coenzyme is specialized to perform one of a small number of biochemical functions but may perform that same function for more than one enzyme. Biotin, for example, serves as the coenzyme for several enzymes, eg, pyruvate carboxylase and acetyl-CoA carboxylase, that catalyze the fixation of CO_2 (Fig 5–8).

Many compounds that serve as coenzymes are either vitamins—low-molecular-weight organic molecules that must be present in small amounts in the human diet to maintain good health—or their metabolites. Biotin, for example, is one of the 9 water-soluble vitamins required by humans. Vitamins are required in the diet because they perform essential functions and yet

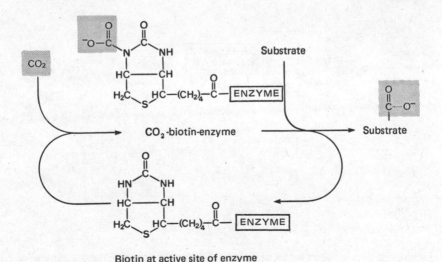

Biotin at active site of enzyme

Figure 5–8. Biotin is the coenzyme for several enzymes that catalyze carboxylation reactions.

either are not synthesized by humans or are produced in amounts insufficient to support normal metabolism.

Regulation of Enzymatic Activity

While enzymes are essential to the metabolism of all organisms, their activities must be regulated to ensure that they act only when needed. There are 4 mechanisms by which the activity of an enzyme can be regulated.

(1) **Allosteric regulation:** An allosteric enzyme is one whose activity is regulated by a compound (an allosteric effector) that binds reversibly to the enzyme at a site other than the active site. An allosteric effector regulates the activity of the enzyme to which it binds by stabilizing a particular conformation of the enzyme. Because an allosteric effector binds at a site other than the active site, it need not resemble the substrates of the enzyme.

The features of allosteric regulation can be illustrated by a description of the bacterial pathway for isoleucine synthesis. In bacteria, threonine is converted to isoleucine by the sequential action of 5 enzymes (Fig 5–9). The activity of this metabolic pathway is controlled by allosteric regulation of the first enzyme of the pathway, threonine deaminase. Isoleucine binds to this enzyme at a site distinct from the active site (Fig 5–10). Upon binding the effector, the enzyme undergoes a conformational change that alters the active site and thereby reduces its activity. Allosteric control of threonine deaminase ensures that when isoleucine accumulates in excess of the needs of the cell, its rate of synthesis will be reduced.

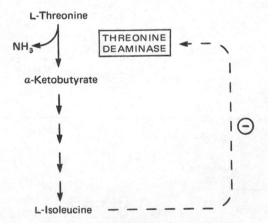

Figure 5–9. L-Isoleucine regulates its own synthesis by controlling the activity of threonine deaminase. Threonine deaminase, the first of 5 enzymes in the pathway that converts threonine to isoleucine, is inhibited by isoleucine, the end product of the pathway.

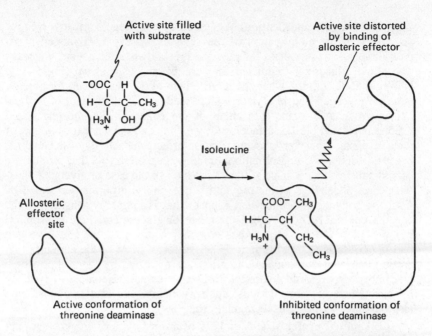

Figure 5–10. Isoleucine regulates the activity of threonine deaminase by changing the conformation of the active site. The jagged arrow denotes the transmission of a conformational change from the site at which the allosteric effector binds to the active site.

The inhibition of threonine deaminase by isoleucine is an example of **feedback inhibition,** ie, inhibition of the activity of an enzyme early in a pathway by the end product of the pathway. Generally, the enzyme subject to feedback inhibition is one that catalyzes the first functionally irreversible step *unique* to the pathway. This reaction is commonly referred to as the **committed step.** In a branched pathway, it is usually the first enzyme after the branch that is regulated.

(2) **Covalent modification:** A second type of regulation involves reversible covalent modification of the regulated enzyme. Glycogen phosphorylase, an enzyme that catalyzes the breakdown of glycogen, a storage polymer of glucose, is regulated, in part, by this mechanism (Fig 5–11). During muscle activity, when the requirement of the muscle cell for glucose is large, glycogen phosphorylase is activated by the attachment of a phosphoryl group to a serine side chain of the enzyme. The phosphorylated residue is located on the surface of the enzyme and does not form a part of the active site. However, phosphorylation affects the activity of the enzyme because it induces a conformational change in the enzyme that produces a catalytically effective shape at the active site. In the case of glycogen phosphorylase, attachment of the regulatory phosphoryl group is performed by a second enzyme, phosphorylase kinase, that is indirectly sensitive to the concentration of glucose. When the need for glucose ends, the phosphoryl group is removed by yet another enzyme, a phosphatase, and glycogen phosphorylase is restored to its less active form. Most of the enzymes that are regulated by covalent modification are modified by the addition of a phosphoryl group to the hydroxyl group of a serine, threonine, or tyrosine residue. However, other groups, such as methyl and acetyl groups, may also be added to enzymes to regulate their activities.

(3) **Limited proteolysis:** Members of a third class of enzymes are activated by cleavage of their polypeptide chains (Fig 5–12). Enzymes of this class are synthesized in inactive forms (proenzymes, or zymogens) and are activated by the proteolytic removal of a short fragment from the amino

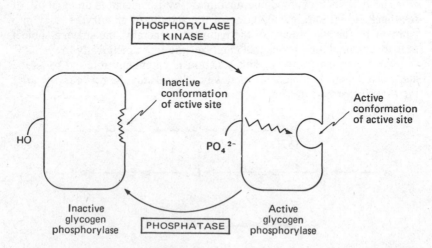

Figure 5–11. Glycogen phosphorylase is regulated by reversible covalent modification. The jagged arrow denotes the transmission of a conformational change from the site of covalent modification to the active site.

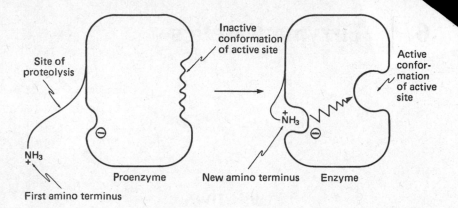

Figure 5–12. Zymogens are activated by limited hydrolysis. The jagged arrow denotes the transmission of a conformational change from the region of the amino terminus to the active site.

terminus. Included in this class are a number of digestive enzymes such as trypsin and chymotrypsin. When the proenzyme form of trypsin—protrypsin—is secreted into the intestine, a 6-amino-acid fragment is removed from the amino terminus. The new amino terminus that is produced by this reaction folds back inside the enzyme and induces a conformational change that activates the enzyme. Unlike the preceding 2 forms of regulation, proteolytic activation is irreversible.

(4) **Control of enzyme production and turnover:** The amount of an enzyme made by a cell may be regulated by increasing or decreasing the rate of either its synthesis or degradation. This form of regulation differs from the others in that the total amount of enzyme polypeptide is changed without changing the catalytic properties of individual enzyme molecules. When the amount of an enzyme is increased, it is said to be **induced;** when it is decreased, it is said to be **repressed.**

Enzyme Kinetics

OBJECTIVES

- Know the definitions of K_m and V_{max} and how these entities are determined experimentally for a given enzyme.

- Know the Michaelis-Menten equation and be able to use it to find the velocity of an enzyme-catalyzed reaction, given the V_{max} of the enzyme and the [S].

- Be able to determine the K_m and V_{max} for an enzyme-catalyzed reaction from a plot of v_i versus [S] or a double-reciprocal plot.

- Be able to distinguish competitive and noncompetitive inhibition using data presented in either a plot of v_i versus [S] or a double-reciprocal plot.

ENZYME KINETICS is the study of the rate behavior of enzyme-catalyzed reactions. Kinetic measurements provide an extremely useful biochemical tool because they allow us to estimate the concentration of an enzyme in a biologic sample and to compare its catalytic activity with that of other enzymes. Kinetic measurements also provide a means of quantitatively describing the effect of a poison or drug on the activity of an enzyme.

The rate at which an enzymatic reaction proceeds is governed in part by the concentration of the enzyme and the concentration of its substrate. This chapter shows first how these variables affect the rate of a reaction and then how the development of enzyme kinetics has provided 2 constructs useful for the description and comparison of enzymes, namely, V_{max} and K_m.

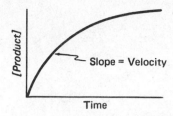

Figure 6–1. The rate of an enzyme-catalyzed reaction corresponds to the increase in product with time.

Initial Velocity

The rate at which a reaction proceeds is measured as the decrease in the concentration of reactants or the increase in the concentration of the products with time. If an enzyme is incubated with its substrate and the appearance of the product with time is recorded on a graph, the resulting line will have, in the simple case, the hyperbolic shape shown in Fig 6–1. The rate of the reaction, which corresponds to the slope of this line, is initially constant but gradually decreases. The decline in the rate of the reaction may be due to depletion of the substrate, inhibition of the enzyme by its product, or denaturation of the enzyme. Each of these events affects the conditions of the reaction, some to an unknown extent. Thus, only in the initial portion of the reaction are the conditions accurately known. For this reason, only the initial velocity (v_i) is used in calculating the kinetic parameters of the reaction. The units of velocity are concentration per unit time, eg, micromoles per minute.

Effects of [S] & [E]

By varying the amount of substrate incubated with a constant amount of enzyme, we can show that the rate of an enzyme-catalyzed reaction depends on the substrate concentration, [S]. For many enzymes, the curve that relates v_i to [S] is hyperbolic (Fig 6–2A). At low substrate concentrations, the rate of the reaction increases almost linearly with increases in the substrate concentration. At high substrate concentrations, the rate reaches a limiting (maximal) value known as V_{max}. The rate of the reaction does not increase indefinitely as the substrate concentration is increased, because the rate of the reaction is also limited by the concentration of the enzyme in the assay. The reaction reaches its V_{max} when the amount of enzyme being assayed is saturated with substrate, and further increases in substrate do not significantly increase the occupancy of the active site.

This observation provides a tool with which to compare the amounts of a given enzyme in several samples. When the concentration of the substrate

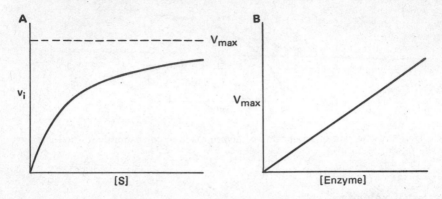

Figure 6–2. The velocity of an enzyme-catalyzed reaction depends on both the substrate and enzyme concentrations. **A:** When the amount of enzyme is constant, the v_i increases as the substrate increases. **B:** When the enzyme is saturated with substrate, the V_{max} of the reaction increases with the amount of enzyme.

is not limiting (ie, when we are measuring the V_{max}), the rate of the reaction is proportionate to the amount of enzyme assayed (Fig 6–2B). The amount of an enzyme in a biologic sample is reported as units of activity. A unit of enzyme is that amount which catalyzes the transformation of 1 μmol of substrate into product in 1 minute.

Michaelis-Menten Equation

Often, it is useful to be able to predict the activity of an enzyme at a given concentration of substrate. If, for example, we can predict the activity of an enzyme throughout the range of substrate concentrations it normally encounters in the body, we can better understand the physiologic role that enzyme plays. The Michaelis-Menten equation, which describes the dependence of reaction velocity on substrate concentration, is used for this purpose.

The basis for the Michaelis-Menten equation is a simple model of enzyme action. Michaelis and Menten proposed that in an enzyme-catalyzed reaction the enzyme combines with its substrate to form an enzyme-substrate complex (ES), which then breaks down either to enzyme and substrate or to enzyme and product.

$$E + S \underset{k_{-1}}{\overset{k_1}{\rightleftharpoons}} ES \overset{k_2}{\rightarrow} E + P$$

The rates at which the partial reactions of this model occur are described by the rate constants k_1, k_{-1}, and k_2. According to this model, the increase in v_i observed with an increase in [S] is due to an increase in the amount of ES formed. At V_{max}, all of the enzyme is involved in an ES complex.

Using this model, Michaelis and Menten developed an equation (shown immediately below) that expresses the initial velocity (v_i) of a reaction in terms of the V_{max}, [S], and the rate constants k_1, k_{-1}, and k_2.

$$v_i = \frac{V_{max}\,[S]}{[S] + \dfrac{(k_{-1} + k_2)}{k_1}}$$

If the ratio of the rate constants is defined as the Michaelis constant (K_m),

$$K_m = \frac{k_{-1} + k_2}{k_1}$$

the Michaelis-Menten equation can then be rewritten as follows:

$$v_i = \frac{V_{max}\,[S]}{[S] + K_m}$$

What does the value of the K_m mean? We can find out by solving the Michaelis-Menten equation for several concentrations of substrate.

$$\text{When } [S] = K_m,\ v_i = \frac{V_{max}}{2}$$

That is to say, K_m is that concentration of substrate at which the rate of the enzyme-catalyzed reaction is half-maximal. (See Fig 6–3.)

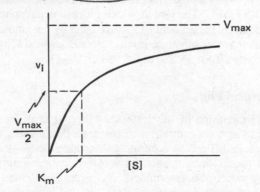

Figure 6–3. The K_m is the concentration of substrate at which the velocity of the reaction is half-maximal.

$$\text{When } [S] = \frac{1}{10} K_m, v_i = \frac{V_{max}}{11}$$

and

$$\text{when } [S] = 10 \times K_m, v_i = \frac{V_{max}}{1.1}$$

In other words, at a concentration of substrate one order of magnitude below the K_m, the velocity of the reaction is only 9% of the maximal velocity, and at a substrate concentration one order of magnitude above the K_m, the enzyme is operating at a velocity near its V_{max}.

As the concentration of the substrate is increased, the rate of an enzymatic reaction also increases. However, 2 enzymes that differ in their K_m values respond to the same increase in substrate concentration with a different increase in rate. This point is illustrated by 2 enzymes important in carbohydrate metabolism, glucokinase and hexokinase. Both enzymes catalyze the transfer of a phosphoryl group from the nucleotide adenosine triphosphate (ATP) to glucose, a sugar.

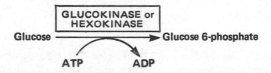

Hexokinase, which catalyzes this reaction in the brain, has a K_m for glucose of 5×10^{-5} mol/L. Glucokinase, which predominates in the liver, has a K_m for glucose of 2×10^{-2} mol/L. Normal blood glucose levels range from 3×10^{-3} to 7×10^{-3} mol/L. Throughout this range of concentration, hexokinase operates virtually at its V_{max}, the velocity increasing only from 98.4 to 99.3% of V_{max} as the glucose concentration is increased from 3×10^{-3} to 7×10^{-3} mol/L. The rate at which glucose is phosphorylated in the brain is limited not by the amount of substrate but by the amount of enzyme. For the same increase in substrate levels, glucokinase exhibits a substantial increase in activity (from 13.2% to 25.9% of V_{max}).

Double-Reciprocal Plots

Owing to limitations in substrate solubility, it is often impossible to assay an enzyme at a concentration of substrate that saturates the enzyme. In such a case, it is difficult to accurately determine the V_{max} of the enzyme from a plot of the velocity versus the substrate concentration, and consequently the K_m cannot be determined. By using an alternative method of plotting kinetic data (the Lineweaver-Burk, or double-reciprocal, plot), it is possible to overcome these difficulties and assess V_{max} and K_m even when only limited kinetic data are available.

While the Michaelis-Menten equation describes a curve, its reciprocal rearranges to the equation for a straight line.

$$\frac{1}{v_i} = \frac{K_m}{V_{max}} \times \frac{1}{[S]} + \frac{1}{V_{max}}$$

In a Lineweaver-Burk plot, the y-axis is $1/v_i$; and the x-axis is $1/[S]$. The y-intercept equals $1/V_{max}$; the x-intercept is $-1/K_m$; and the slope of the line is K_m/V_{max} (Fig 6–4).

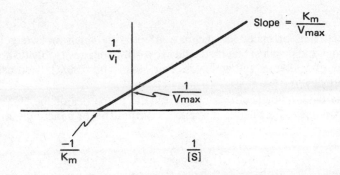

Figure 6–4. A Lineweaver-Burk plot of the kinetic data shown in Fig 6–3.

Enzymes Not Displaying Michaelis-Menten Kinetics

For many enzymes, including most allosteric enzymes, a plot of v_i versus [S] is not hyperbolic. In these cases, it may be either a sigmoid curve or a flattened hyperbola (Fig 6– 5). Many enzymes that display these kinetic

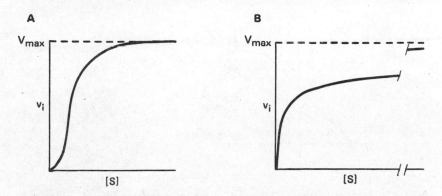

Figure 6–5. A: A sigmoid dependence of v_i on [S] indicates positive cooperativity. **B:** A flattened hyperbola suggests negative cooperativity.

properties are multimers that have more than one active site. Their kinetic behavior indicates cooperativity between the subunits of the multimer. An enzyme whose rate behavior is described by a sigmoid curve is very sensitive to substrate concentration and responds to a small increase in substrate with a large increase in activity. The sigmoid shape of the curve indicates positive cooperativity in the catalytic event; ie, activity at one active site increases the chances of activity at other active sites in the same enzyme. If the curve is a flattened hyperbola, the enzyme is relatively insensitive to changes in substrate concentration. This behavior may indicate negative cooperativity.

Inhibitors

The activities of many enzymes are affected by small molecules, including drugs and poisons, that bind either reversibly or irreversibly to them. The effects of inhibitors and activators can readily be characterized using enzyme kinetics.

On the basis of their effects on the kinetic properties of the enzyme, reversible inhibitors are classified as either **competitive** or **noncompetitive.**

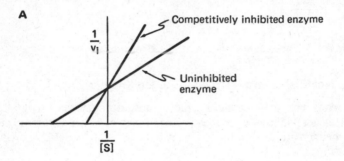

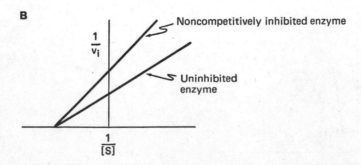

Figure 6–6. *A:* A competitive inhibitor increases the apparent K_m of an enzyme-catalyzed reaction but does not alter the V_{max}. *B:* A noncompetitive inhibitor decreases the V_{max} of the reaction but does not change the K_m.

Competitive inhibitors, as the name implies, compete with t...
the enzyme; ie, they reduce the probability that the substra...
with the enzyme to form ES. Therefore, competitive inhi...
apparent K_m of the enzyme for the substrate (Fig 6 – 6A).
inhibitor has no effect on the V_{max} of a reaction because at infinite substrate
concentration, the condition under which V_{max} is measured, the substrate
saturates the enzyme, excluding the competitive inhibitor.

In contrast, noncompetitive inhibitors do not compete with the substrate
of the reaction for occupancy of the enzyme and therefore do not prevent
formation of the enzyme-substrate complex. Instead, they reduce the rate
at which the ES productively dissociates to yield the product of the reaction.
Noncompetitive inhibitors may either totally prevent or retard this step of
the reaction. A noncompetitive inhibitor decreases the V_{max} of the reaction
it catalyzes while having no effect on the K_m (Fig 6 – 6B). The noncompetitive
inhibitor thus has the effect of simply removing some of the enzyme molecules from the reaction.

Section II: Metabolism

7 | # Bioenergetics &
    ## Fuel Metabolism

OBJECTIVES

- Know what is meant by a high-energy bond and be able to recognize compounds in which one occurs.

- Know the definition of an oxidation-reduction reaction. Learn the names and be able to recognize the structures of the major coenzymes of oxidation-reduction reactions.

- Be able to describe the reactions of glycolysis. Learn the names of the intermediates and enzymes of the pathway. Know which reactions are thermodynamically irreversible. Be able to list the products of aerobic and anaerobic glycolysis.

- Be able to explain how fructose and galactose enter the glycolytic pathway.

- Be able to explain the purpose and the operation of the Cori cycle.

- Be able to describe the steps in the reaction catalyzed by pyruvate dehydrogenase. Know the names, origins, and functions of the coenzymes that participate in this reaction.

- Be able to describe the transport of fatty acids into the mitochondria and their catabolism via β-oxidation.

- Know the names and physiologic roles of ketone bodies. Be able to describe their synthesis from hepatic fatty acids.

- Be able to describe the reactions of the citric acid cy
  names of the intermediates and the names and intrace
  of the enzymes involved. Know how carbon atoms e
  the cycle, which reactions involve oxidation of a substr.
  reaction involves substrate level phosphorylation.

- Be able to explain how the electron transport chain and mitochondrial
  ATPase participate in ATP production.

- Be able to calculate the yield of ATP for the complete oxidation of
  glucose and for the complete oxidation of palmitate.

- Be able to explain how glycolysis, the citric acid cycle, and oxidative
  phosphorylation are regulated by the size of the ATP pools and by
  oxygen availability.

ONLY THOSE REACTIONS that involve a decrease in free energy, ie, those that
have a negative ΔG, proceed spontaneously (see Chapter 1). Yet many
biologically important processes, such as the synthesis of macromolecules,
the contraction of muscles, and the concentration of molecules against chemical gradients, would not by themselves be energetically favorable. Energy
to drive these reactions must come from another source. In biology, energy-
requiring reactions are driven by being coupled chemically to other reactions
that are energetically favorable. The free energy needed to drive unfavorable
reactions is provided by the metabolism of fuels—carbohydrates, lipids
(fats), and amino acids. Each of these compounds can be **catabolized** (broken
down) in pathways that extract energy and make it available to drive other
reactions. In this chapter, we will survey the pathways of fuel catabolism.

High-Energy Compounds

The free energy made available through the catabolism of fuels is not
transmitted directly to energy-requiring reactions but instead is used to synthesize a compound that acts as a carrier of free energy, **adenosine triphosphate (ATP)**. ATP (Fig 7–1) is a member of a class of compounds
called nucleoside triphosphates (see Chapter 12). It has 3 component parts:
a base (adenine), a sugar (ribose), and 3 phosphoryl groups that are joined
to ribose by a phosphate ester bond and to each other by phosphoanhydride
bonds.

From the perspective of fuel metabolism, the most important parts of
the ATP molecule are its 2 **phosphoanhydride bonds**. Breakdown of either
of these bonds is accompanied by a large decrease in free energy; ie, breakdown is a highly favorable reaction. For example, the $\Delta G^{0'}$ for hydrolysis

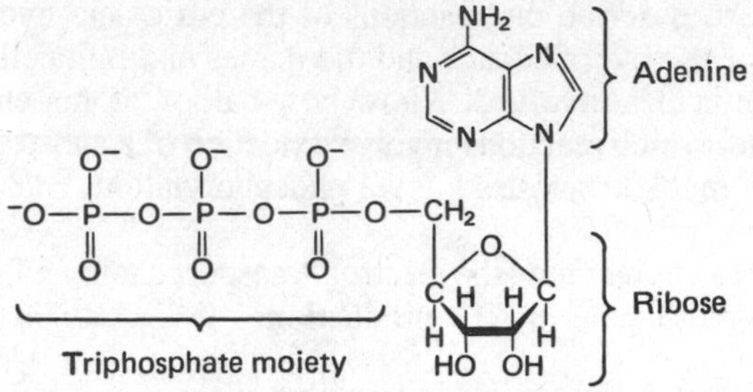

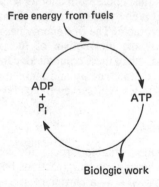

Figure 7–1. The structure of ATP.

of ATP to either ADP (adenosine diphosphate) and P_i (inorganic phosphate) or AMP (adenosine monophosphate) and PP_i (pyrophosphate) is −7.3 kcal/mol. Any bond whose breakdown is accompanied by a large decrease in free energy (≥ 5 kcal/mol) is termed a **high-energy bond,** represented by the symbol ∼. Because ATP has high-energy bonds, it is able to serve as a link between energy-yielding and energy-requiring processes (Fig 7–2). The phosphoanhydride bonds of ATP are formed at the expense of energy-yielding reactions of catabolism. ATP passes on that free energy to energy-requiring processes in reactions that break the phosphoanhydride bonds.

Free energy from fuels

ADP
+
P_i

ATP

Biologic work

Figure 7–2. ATP is the link between energy-yielding and energy-consuming biologic processes.

Other compounds, such as **phosphoguanidines, phosphoric carboxylic acid anhydrides, thioesters, enol phosphates,** and **nucleoside triphosphates** other than ATP, also contain high-energy bonds (Fig 7–3). The free energy of these compounds is used in a number of reactions that directly or indirectly lead to ATP production.

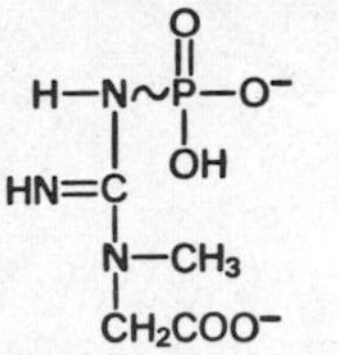

$$
\begin{array}{c}
\text{O} \\
\| \\
\text{H—N} \sim \text{P—O}^- \\
\text{OH} \\
\text{HN=C} \\
\text{N—CH}_3 \\
\text{CH}_2\text{COO}^-
\end{array}
$$

Creatine phosphate

A phosphoguanidine

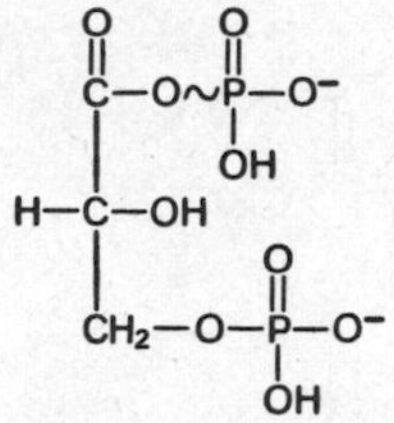

$$
\begin{array}{c}
\text{O} \quad \text{O} \\
\| \quad \| \\
\text{C—O} \sim \text{P—O}^- \\
\text{OH} \\
\text{H—C—OH} \\
\text{O} \\
\| \\
\text{CH}_2\text{—O—P—O}^- \\
\text{OH}
\end{array}
$$

1,3-Bisphosphoglycerate

A phosphoric carboxylic acid anhydride

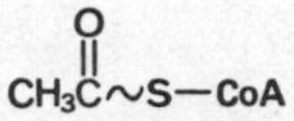

$$
\begin{array}{c}
\text{O} \\
\| \\
\text{CH}_3\text{C} \sim \text{S—CoA}
\end{array}
$$

Acetyl-CoA

A thioester

$$
\begin{array}{c}
\text{CH}_2 \quad \text{O} \\
\| \quad \| \\
\text{C—O} \sim \text{P—O}^- \\
\text{COO}^- \quad \text{OH}
\end{array}
$$

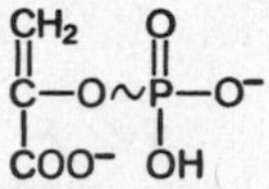

Phosphoenolpyruvate

An enol phosphate

Figure 7–3. Types of compounds that contain high-energy bonds. In each compound, the high-energy bond is represented by $\sim$.

Oxidation-Reduction Reactions

Catabolic pathways include a number of oxidation-reduction reactions, ie, reactions in which electrons are transferred from one compound to another. The compound that loses the electrons is said to be oxidized, and that which gains the electrons is said to be reduced. Electrons can be transferred alone (as e^-), as part of a hydrogen atom ($H^+ + e^-$), or as part of a hydride ion, H^- ($H^+ + 2e^-$). An electron transferred in any of these forms is referred to as a **reducing equivalent.**

In aerobic organisms, oxygen is the ultimate acceptor of electrons. Fuels are not, however, directly oxidized by O_2. Instead, they transfer electrons to 3 specialized electron-carrying coenzymes: **nicotinamide adenine dinucleotide (NAD), nicotinamide adenine dinucleotide phosphate (NADP),** and **flavin adenine dinucleotide (FAD)** (Fig 7–4). In humans, NAD and NADP are formed from the vitamin **niacin** and FAD from **riboflavin** (vitamin B_2). NAD and NADP each accept a hydride ion to become NADH and NADPH, respectively; FAD accepts 2 H atoms, forming $FADH_2$. The reducing equivalents carried by NAD and FAD are subsequently used in the synthesis of ATP; those of NADPH do not contribute to ATP production and are instead reserved for biosynthetic (anabolic) reactions (Fig 7–5).

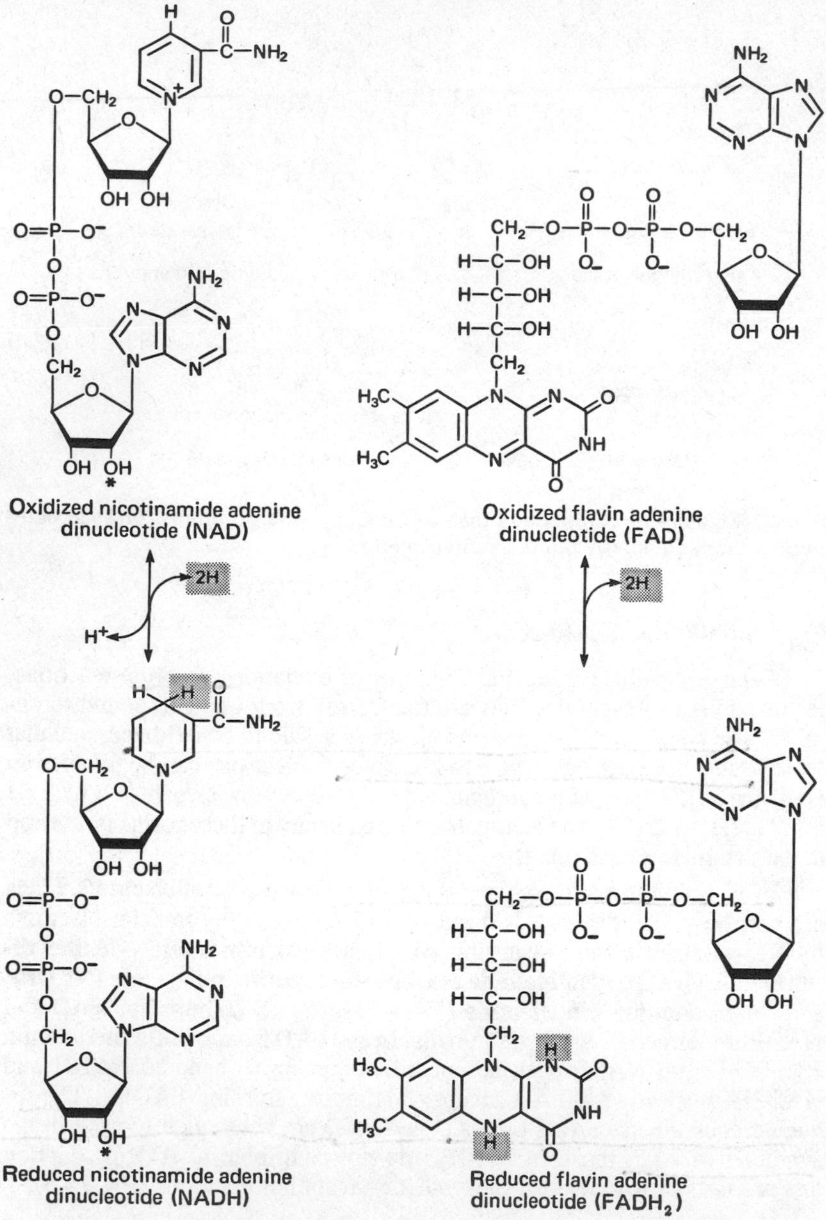

Figure 7–4. NAD, NADP, and FAD are carriers of reducing equivalents. The structure of NADP differs from that of NAD only by addition of a phosphate at the position marked *. In their oxidized forms, the pyridine rings of NAD and NADP both carry a positive charge. To emphasize this point, these compounds are represented as NAD^+ and $NADP^+$ in figures. The positive charge is omitted in the text.

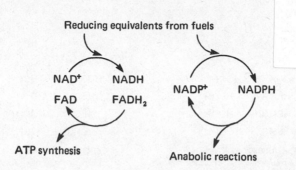

Figure 7–5. Reducing equivalents from fuels are used in ATP synthesis and in anabolic reactions.

Metabolic Fuels

Carbohydrates, lipids, and the carbon skeletons of amino acids all serve as fuels for human metabolism.

Carbohydrates, also called saccharides, are polyhydroxyaldehydes, polyhydroxyketones, and their derivatives. Carbohydrates that contain one aldehyde or ketone group are termed **monosaccharides.** These compounds are unbranched chains that range in length from 3 carbons (trioses) to 8 carbons (octoses). Carbohydrates containing more than 8 carbons are made by joining multiple monosaccharide units to form **disaccharides** (2 sugar units), **oligosaccharides** (3–6 units), and **polysaccharides** (more than 6 units).

The structural features of monosaccharides are illustrated in Fig 7–6. If the carbonyl group is at the end of the chain, the sugar is termed an **aldose;** if not, the sugar is a **ketose** (Fig 7–6A). Most monosaccharides have one or more asymmetrically substituted carbon atoms and therefore exhibit stereoisomerism (see Chapter 1). Two sugars that are mirror images of each other **(enantiomers)** share the same common name, eg, glucose, and are distinguished by the designation D or L (Fig 7–6B). For a monosaccharide with 2 or more asymmetric carbon atoms, the designation D or L is based on the configuration of the substituents around the asymmetric carbon atom most distant from the carbonyl group. Only the D isomers of sugars participate in fuel metabolism. Sugars that differ only in the configuration around one carbon atom are **epimers** of each other (Fig 7–6C). In monosaccharides that contain 5 or more carbons, the carbonyl carbon can react with a hydroxyl group near the other end of the chain to form a ring structure (Fig 7–6D). Although the open chain and the ring forms are in equilibrium with each other, the ring form is more stable and predominates. When the sugar adopts the ring conformation, the carbonyl carbon becomes asymmetrically substituted. Thus, the ring has 2 isomeric forms, termed **anomers.** For D sugars, if the hydroxyl group of the anomeric carbon projects

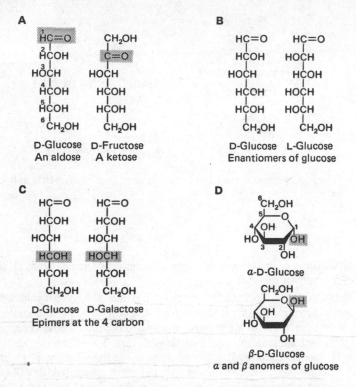

Figure 7–6. Important structural features of monosaccharides. The student should bear in mind that because the formulas shown here do not show the tetrahedral geometry of the carbon atom, they do not accurately represent the 3-dimensional shapes of the carbohydrates depicted.

below the ring, the compound is designated the α anomer; in the β anomer, it projects above the ring. The hydroxyl group attached to the anomeric carbon participates in reactions between monosaccharides and other compounds. Condensation of this group with a hydroxyl group of another compound results in the formation of a glycosidic linkage.

The most abundant carbohydrates in the human diet are the monosaccharides glucose and fructose; the disaccharides sucrose (glucose plus fructose), lactose (glucose plus galactose), and maltose (2 glucose units); and the polysaccharides starch (an α glycoside) and cellulose (a β glycoside). Because only monosaccharides can be absorbed from the intestine, larger carbohydrates must first be degraded by digestive enzymes. Humans do not have an enzyme that can break the β-glycosidic linkage of the cellulose polymer.

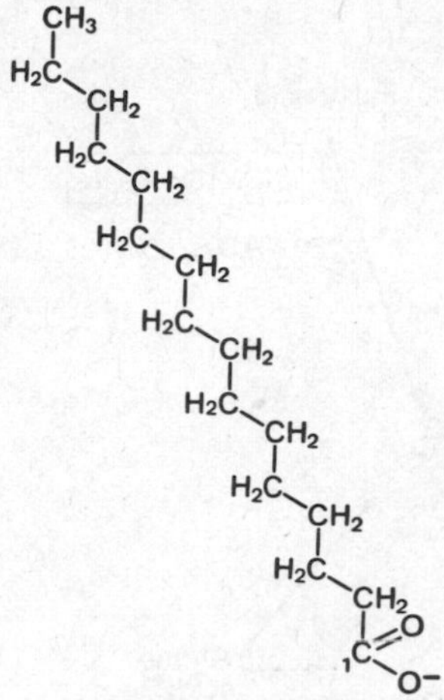

Palmitate
(16:0)

A saturated fatty acid

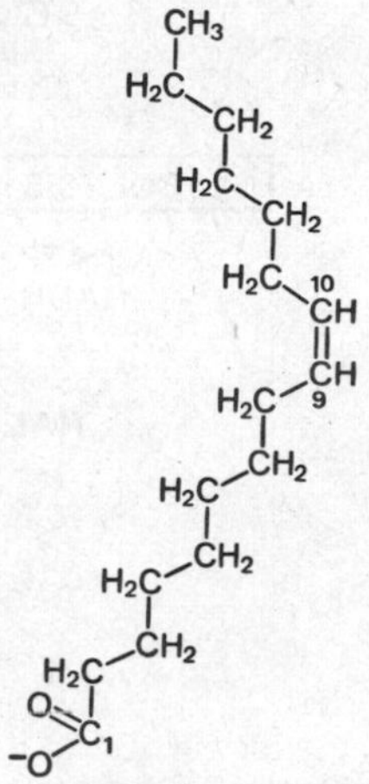

Palmitoleate (*cis* form)
(16:1;9)

An unsaturated fatty acid

Figure 7–7. A saturated fatty acid, palmitate (16:0), and an unsaturated fatty acid, palmitoleate (16:1;9).

Lipids are biologic compounds that are more soluble in nonpolar solvents than in water. The fuel lipids, **fatty acids**, are long, unbranched hydrocarbon chains that terminate in a carboxyl group (Fig 7–7). The most abundant fatty acids contain an even number of carbon atoms and range in length from 14 to 22 carbons. Many fatty acids are unsaturated (ie, they contain one or more double bonds), and most naturally occurring unsaturated fatty acids are *cis* isomers. Several different systems of nomenclature are used to describe fatty acids. In this text, fatty acids will be named using a numerical convention that specifies the number of carbon atoms, the number of double bonds, and their location. The carbon atoms of a fatty acid are numbered starting with the carboxyl carbon. The location of a double bond is described by the number of that carbon atom of the bond nearest the carboxyl group. In this system, palmitoleate, the 16-carbon fatty acid that contains a double bond between carbons 9 and 10, is designated 16:1;9.

Catabolic Pathways

Through the operation of a small number of catabolic pathways, carbohydrates, lipids, and amino acids are oxidized to CO_2 and H_2O, and the free energy released in the process is used to phosphorylate ADP to ATP (Fig 7–8). The catabolic pathways also provide substrates for many of the anabolic (synthetic) pathways of the cell.

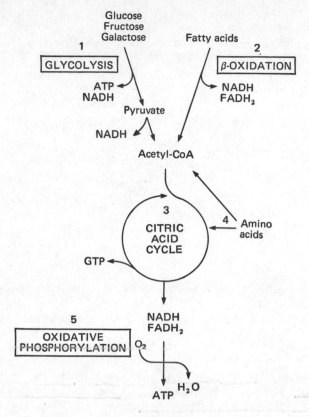

Figure 7–8. An overview of fuel catabolism.

(1) The first steps in the catabolism of the monosaccharides glucose, fructose, and galactose are carried out by the enzymes of **glycolysis.** The glycolytic pathway converts these sugars to pyruvate, a 3-carbon acid, and uses the energy released in the process to phosphorylate ADP to ATP and reduce NAD to NADH. Pyruvate is further metabolized to an activated 2-carbon fragment, acetyl-CoA, in a process that generates more NADH.

(2) Fatty acids are catabolized by the β-**oxidation** pathway. The products of β-oxidation are acetyl-CoA, NADH, and $FADH_2$.

(3) The acetyl-CoA produced by these pathways is oxidized to CO_2 by a series of enzymes that together form the **citric acid cycle.** In addition to CO_2, this cycle produces NADH, $FADH_2$, and a high-energy compound, guanosine triphosphate (GTP).

(4) **Amino acid catabolism** is initiated by a series of pathways in which amino acids are deaminated and converted to acetyl-CoA, pyruvate, and

intermediates of the citric acid cycle. The deamination of amino acids is discussed in Chapter 10.

(5) The NADH and FADH$_2$ produced through the operation of glycolysis, β-oxidation, and the citric acid cycle are reoxidized by donating their electrons to an electron transport chain which, in turn, tranfers the electrons to O$_2$. The operation of the electron transport chain is energetically linked to mitochondrial ATP synthesis. The coupled process of electron transport and ATP synthesis is termed **oxidative phosphorylation.**

Glycolysis

The initial steps in the catabolism of glucose are carried out by the enzymes of glycolysis, which are located in the cytoplasm. All human tissues contain glycolytic enzymes and are therefore able to metabolize glucose. The overall reaction catalyzed by the pathway is

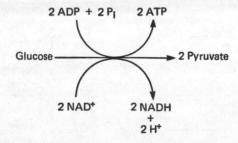

The individual reactions of glycolysis are shown in Fig 7–9.

Glycolysis can be divided into 2 phases: an ATP-consuming phase and an ATP-producing phase. In the first phase, the ATP is used to convert 1 molecule of glucose to 2 molecules of a phosphorylated 3-carbon sugar. First, a phosphoryl group is transferred from ATP to glucose to form glucose 6-phosphate (reaction 1). This reaction can be catalyzed by 2 different enzymes, hexokinase and glucokinase. Hexokinase, which has a low K_m for glucose (eg, for brain hexokinase, 5×10^{-5} mol/L), is found in all tissues and acts to keep the cell supplied with intermediates of glycolysis. Glucokinase, which has a much higher K_m for glucose (2×10^{-2} mol/L), predominates in the liver. As shown in Chapter 8, glucokinase serves primarily to direct excess glucose into storage compounds and thus helps to regulate blood glucose levels.

Glucose 6-phosphate is isomerized to fructose 6-phosphate (reaction 2), which is then phosphorylated at the expense of a second ATP to form fructose 1,6-bisphosphate (reaction 3). The enzyme that catalyzes this second phosphorylation, phosphofructokinase, is the rate-limiting enzyme in glycolysis. Fructose 1,6-bisphosphate is cleaved to 2 triose phosphates, glyceraldehyde 3-phosphate and dihydroxyacetone phosphate (reaction 4). The latter is isomerized to glyceraldehyde 3-phosphate (reaction 5).

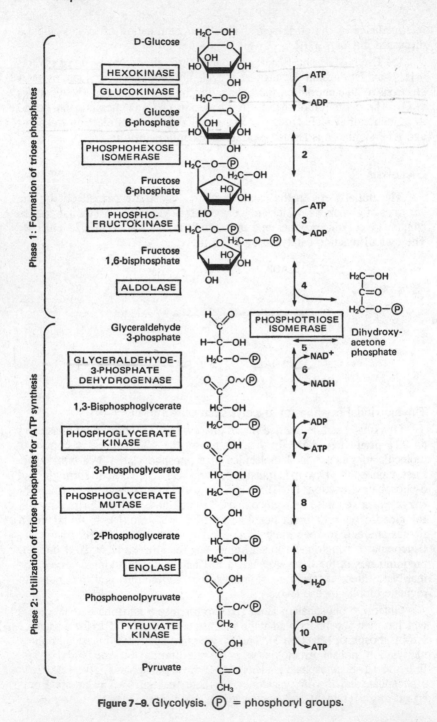

Figure 7–9. Glycolysis. ⓟ = phosphoryl groups.

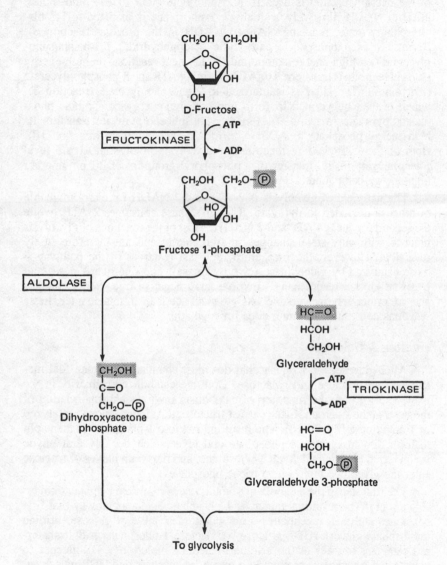

Figure 7–10. Catabolism of fructose. (P) = phosphoryl groups.

In the 5 reactions that make up the second phase of glycolysis, glyceraldehyde 3-phosphate is converted to pyruvate, and the free energy change of the overall reaction is used to phosphorylate ADP to ATP and reduce NAD to NADH. First, glyceraldehyde 3-phosphate is oxidized to 1,3-bisphosphoglycerate, reducing NAD to NADH in the process (reaction 6). Because it is a phosphoric carboxylic acid anhydride, 1,3-bisphosphoglycerate is a high-energy compound. In the next reaction, its high-energy phosphate group is transferred to ADP, forming ATP and 3-phosphoglycerate (reaction 7). The latter is isomerized to 2-phosphoglycerate (reaction 8), which is then dehydrated to form another high-energy compound, phosphoenolpyruvate (reaction 9). Finally, phosphoenolpyruvate transfers its high-energy phosphate to ADP to form ATP and pyruvate (reaction 10). Both of the ATP-yielding reactions of glycolysis involve **substrate level phosphorylation,** ie, transfer of a phosphoryl group to ADP from another high-energy compound.

The net yield of glycolysis is 2 ATP and 2 NADH for every molecule of glucose degraded to pyruvate. The first phase consumes 2 ATP, while the second produces 4 ATP and 2 NADH. Most of the reactions of glycolysis proceed with only small changes in free energy and are therefore freely reversible. However, the first, third, and last reactions of the pathway— those catalyzed by glucokinase (or hexokinase), phosphofructokinase, and pyruvate kinase, respectively—involve large negative free energy changes and are effectively irreversible. As we shall see later in this chapter, these reactions constitute important sites for regulation.

Fructose & Galactose

After glucose, the monosaccharides most abundant in human fuel metabolism are fructose and galactose. Both are catabolized primarily in the liver, and for both the first step of catabolism involves phosphorylation of the sugar at the 1-carbon. In the case of fructose, this first reaction is catalyzed by fructokinase (Fig 7– 10). The resulting fructose 1-phosphate is then split to form dihydroxyacetone phosphate and glyceraldehyde. Glyceraldehyde is converted to glyceraldehyde 3-phosphate, and thus both halves of fructose enter the glycolytic pathway as triose phosphates.

The first step in the catabolism of galactose is catalyzed by galactokinase (Fig 7– 11). The resulting galactose 1-phosphate enters a pathway that first consumes and then regenerates a nucleotide derivative of glucose, uridine diphosphate glucose (UDP-glucose). Galactose-1-phosphate uridyltransferase catalyzes transfer of the uridine nucleotide moiety of UDP-glucose to galactose 1-phosphate, producing glucose 1-phosphate and UDP-galactose. UDP-galactose is then converted to UDP-glucose through the action of an epimerase. UDP-glucose may either serve as a substrate for glycogen synthesis (see Chapter 8) or, via the transferase, be converted to glucose 1-phosphate. The latter may be isomerized to glucose 6-phosphate and thus enter glycolysis.

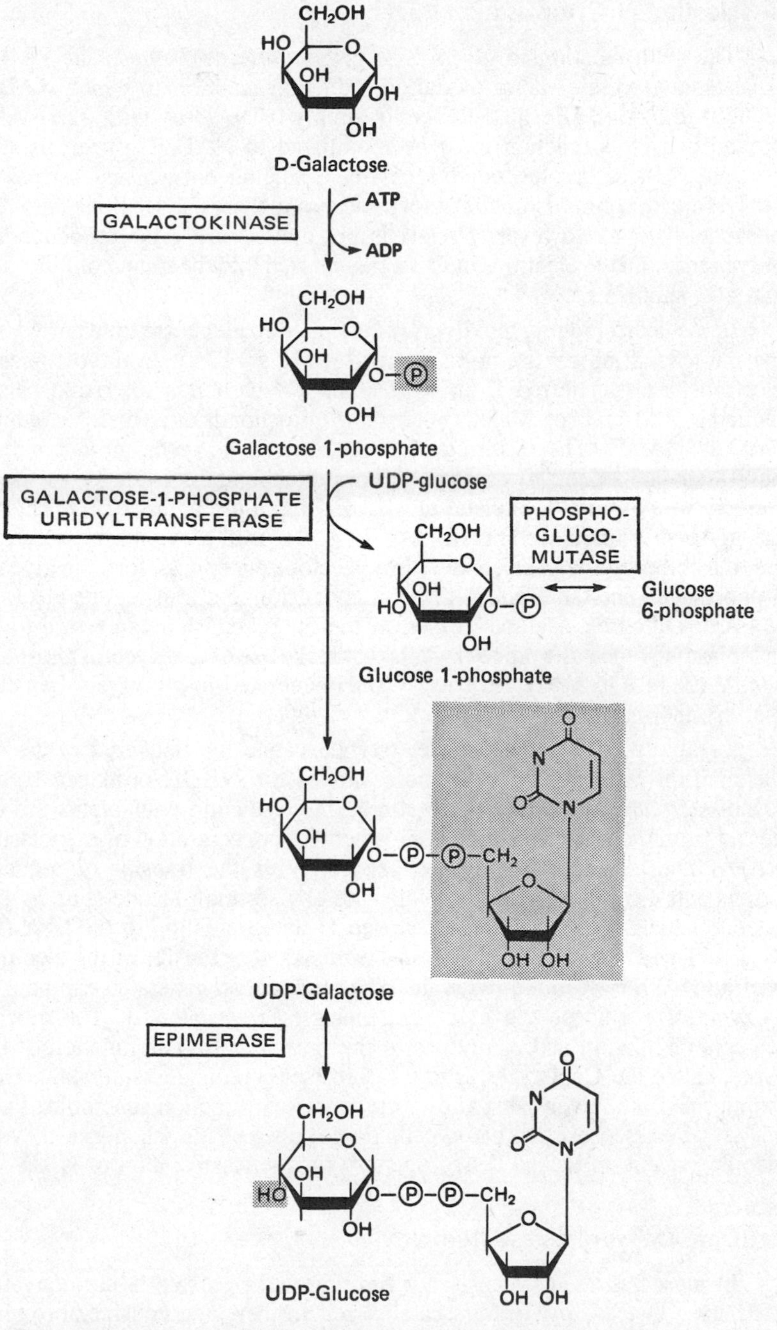

Figure 7–11. Catabolism of galactose. $\textcircled{P}$ = phosphoryl groups.

Reoxidation of Cytoplasmic NADH

The sixth reaction in glycolysis (glyceraldehyde 3-phosphate → 1,3-bisphosphoglycerate) is an oxidation-reduction reaction in which NAD is reduced to NADH. Because the cell's supply of NAD is finite, the NADH produced by this reaction must be reoxidized to NAD if glycolysis is to continue. Under aerobic conditions, the reducing equivalents carried by NADH are transported into the mitochondria and there used to drive oxidative phosphorylation. However, NADH is not able to cross the mitochondrial membrane, and the electrons must be transported into the mitochondria using one of 2 shuttles.

In the heart, kidney, and liver, reducing equivalents are transported into the mitochondrion by the **malate shuttle** (Fig 7–12A). In the cytoplasm, electrons are transferred from NADH to a 4-carbon acid, oxaloacetate, reducing it to malate. Malate enters the mitochondrion, where it reduces NAD to NADH. The resulting oxaloacetate can leave the mitochondrion only after conversion to aspartate. Extramitochondrial oxaloacetate is re-formed from aspartate. Skeletal muscle and brain use an alternative shuttle, the **glycerol phosphate shuttle** (Fig 7–12B). In the cytoplasm, electrons are transferred from NADH to dihydroxyacetone phosphate, forming glycerol 3-phosphate. The latter enters the mitochondrion and transfers its electrons to FAD. Dihydroxyacetone formed in the mitochondrion can return to the cytoplasm to repeat the process. Cells that make use of the glycerol phosphate shuttle produce less ATP for each NADH generated in glycolysis than cells that use the malate shuttle (see p 93).

In rapidly contracting muscle, oxygen cannot be delivered to the mitochondrion fast enough to reoxidize all of the NADH produced by the glycolytic pathway. Under these conditions, reducing equivalents are exported from the muscle to the liver. When the concentration of cytoplasmic NADH rises, **lactate dehydrogenase** catalyzes the transfer of reducing equivalents from NADH to pyruvate, thereby forming lactate (Fig 7–13). Lactate leaves the muscle and is carried in the circulation to the liver (Fig 7–14). There, lactate dehydrogenase catalyzes the transfer of the electrons back to NAD, re-forming pyruvate. The pyruvate so generated can be used to synthesize glucose via gluconeogenesis (see Chapter 8). The glucose reenters circulation and can return to the muscle cell. The function of this cycle, called the **Cori cycle,** is to transfer excess reducing equivalents from the muscle to the liver. This allows the muscle to function anaerobically for a short time. Because red blood cells do not possess mitochondria, they are also dependent on lactate dehydrogenase for the regeneration of NAD.

Pyruvate Dehydrogenase Complex

In most tissues, oxygen is not limiting and pyruvate is not converted to lactate. Instead, pyruvate is catabolized further, and considerably more ATP is produced. In the first step of this process, pyruvate is transported

A. The malate shuttle

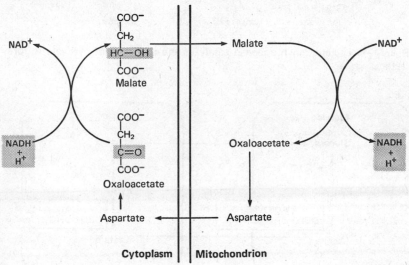

B. The glycerol phosphate shuttle

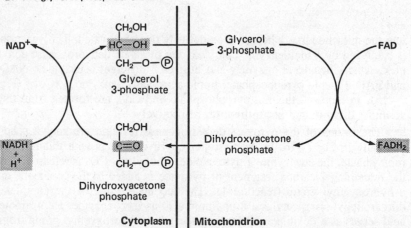

Figure 7–12. *A:* The malate shuttle. *B:* The glycerol phosphate shuttle.

YIELDS 3 ATP   YIELDS 2 ATP

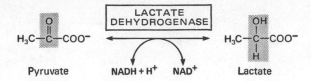

Figure 7–13. Lactate dehydrogenase catalyzes the reduction of pyruvate to form lactate.

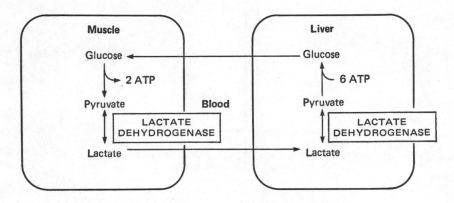

Figure 7–14. The Cori (lactic acid) cycle.

into the mitochondria, where it is oxidatively decarboxylated. This reaction is catalyzed by a multienzyme complex, the pyruvate dehydrogenase complex, which includes 3 enzymes and utilizes 5 different coenzymes: NAD and FAD, thiamin pyrophosphate, lipoic acid, and coenzyme A (CoA) (Fig 7–15). In humans, thiamin pyrophosphate and CoA are formed from the vitamins **thiamin** and **pantothenate,** respectively.

The events of the pyruvate dehydrogenase reaction are shown schematically in Fig 7–16. In the first step, pyruvate reacts with thiamin pyrophosphate, the coenzyme of pyruvate decarboxylase. CO_2 is released, and the remaining 2-carbon fragment of pyruvate is added to the coenzyme as a hydroxyethyl group (reaction 1). The second enzyme of the complex, dihydrolipoyl transacetylase, uses lipoic acid as its coenzyme. Here, lipoic acid serves as a flexible arm that first accepts the hydroxyethyl group from thiamin pyrophosphate (reaction 2) and then donates it, as an acetyl group, to CoA (reaction 3). Lipoic acid is reduced to its dithiol form in the process. Lipoic acid is reoxidized by dihydrolipoyl dehydrogenase and its tightly bound coenzyme, FAD (reaction 4). The reducing equivalents are finally

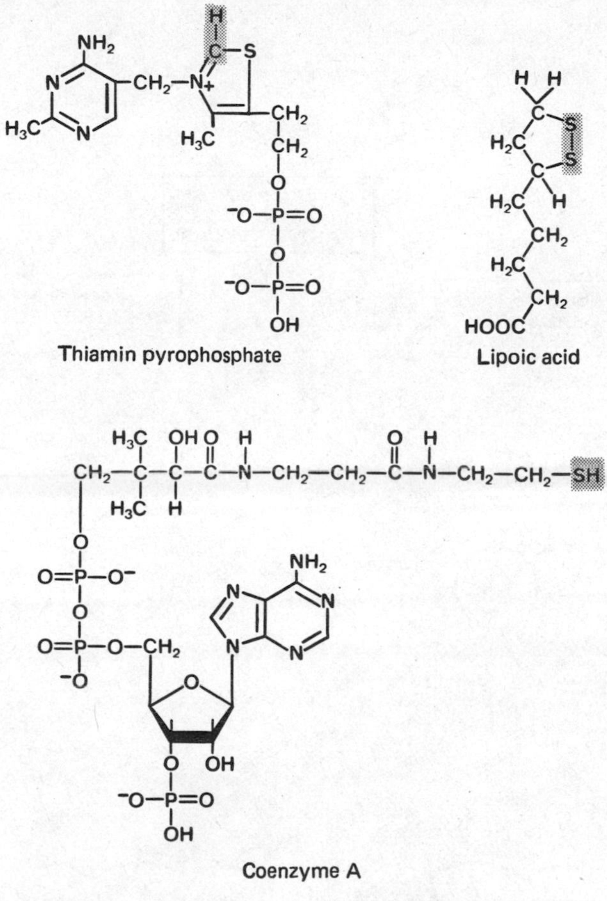

Figure 7–15. The structures of thiamin pyrophosphate (TPP), lipoic acid, and CoA. The reactive group of each coenzyme is shaded.

transferred from $FADH_2$ to a free molecule of NAD, forming NADH (reaction 5). Although no ATP is produced in the reaction catalyzed by the pyruvate dehydrogenase complex, a part of the free energy of pyruvate is captured in the form of acetyl-CoA and NADH. Because acetyl-CoA is a thioester compound, it contains a high-energy bond. The energy of this compound is used when acetyl-CoA is further metabolized in the citric acid cycle.

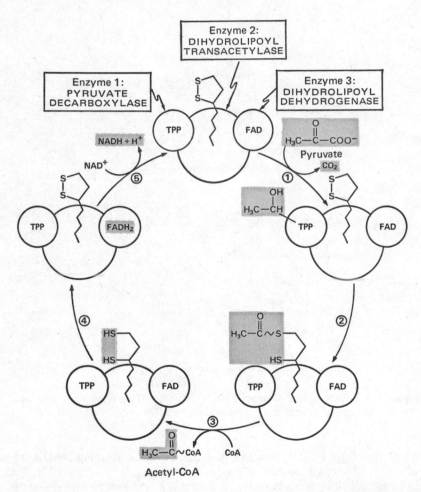

Figure 7–16. Pyruvate dehydrogenase catalyzes a 5-step reaction. TPP = thiamin pyrophosphate.

Fatty Acid Catabolism

The major pathway by which fatty acids are catabolized, **β-oxidation,** is located in the matrix (the interior) of the mitochondria. Thus, cells that lack these organelles, eg, red blood cells, cannot use fatty acids as a fuel. Because the inner membrane of the mitochondrion is impermeable to both free fatty acids and their CoA derivatives, fatty acids must be converted to another form for transport into the mitochondria. Fatty acids are transported into mitochondria attached to a carrier called **carnitine** (Fig 7–17). Cytoplasmic fatty acids are first joined to CoA in an ATP-dependent reaction catalyzed by a **thiokinase.** Fatty acyl-CoA donates its acyl group to carnitine, forming an oxygen ester. Acyl carnitine enters the mitochondrion and transfers the acyl group once more to CoA.

•The enzymes of the β-oxidation pathway catalyze a series of reactions in which 2-carbon units are removed from the carboxyl end of the activated fatty acid chain and are released as acetyl-CoA. This results in a progressive shortening of the fatty acid chain. The 4 steps of this process are shown in Fig 7–18.

In the first reaction, acyl-CoA dehydrogenase oxidizes fatty acyl-CoA. This reaction introduces a double bond between the second and third carbons of the chain, forming the *trans* isomer of the unsaturated fatty acid. The 2 hydrogen atoms given up by the fatty acid are transferred to a molecule of FAD that serves as a tightly bound coenzyme for the dehydrogenase. Next, the double bond is hydrated. The third (β) carbon is then oxidized to form 3-ketoacyl-CoA, while NAD is reduced to NADH. Finally, a free molecule of CoA forms a thioester bond with the newly oxidized carbon, releasing acetyl-CoA from the chain. The overall reaction mediated by the pathway is thus

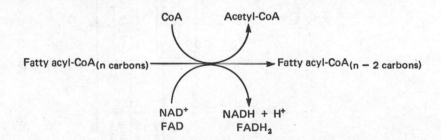

The shortened acyl-CoA can then serve as a substrate for acyl-CoA dehydrogenase, and the cycle is repeated.

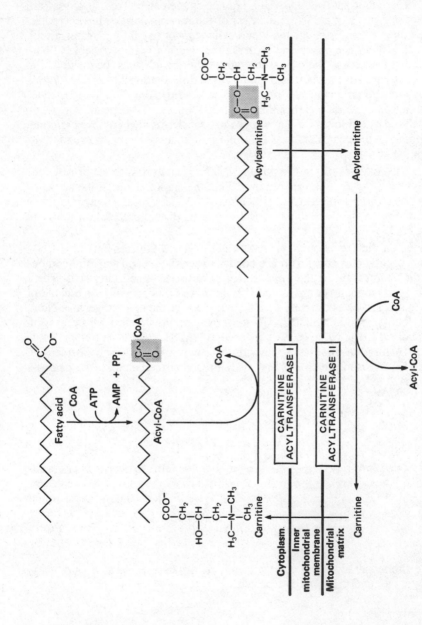

Figure 7–17. Fatty acids are transported into the mitochondria by the carnitine shuttle.

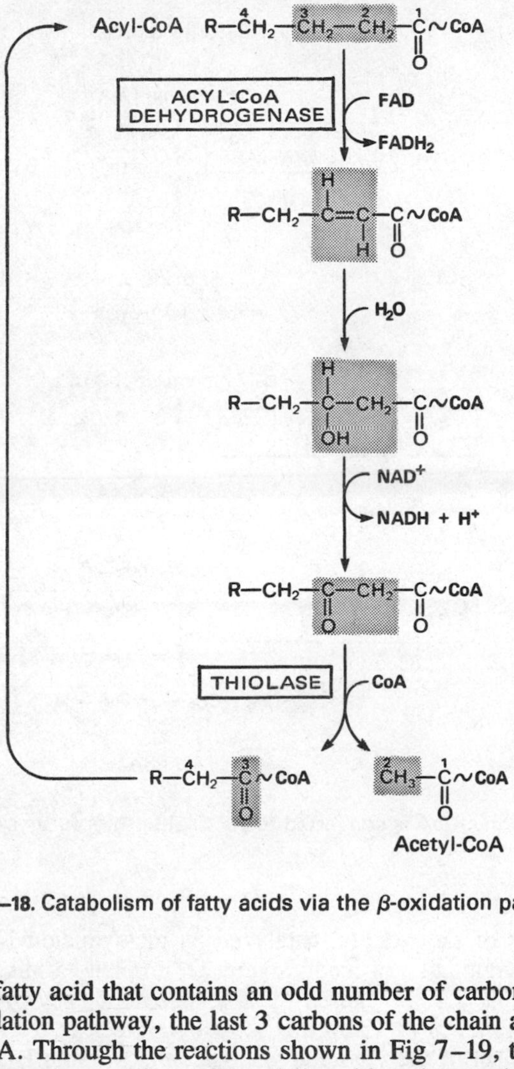

Figure 7–18. Catabolism of fatty acids via the β-oxidation pathway.

   When a fatty acid that contains an odd number of carbons is degraded by the β-oxidation pathway, the last 3 carbons of the chain are released as propionyl-CoA. Through the reactions shown in Fig 7–19, this compound is converted to an intermediate of the citric acid cycle, succinyl-CoA. Two coenzymes participate in the metabolism of the 3-carbon fragment. **Biotin** serves as the coenzyme of propionyl-CoA carboxylase, and **adenosylcobalamin,** a metabolite of vitamin B_{12}, participates in the reaction catalyzed by methylmalonyl-CoA mutase.

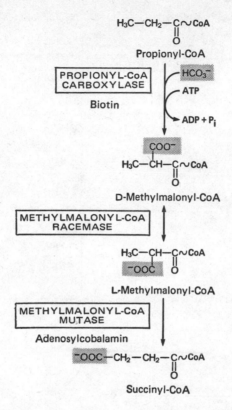

Figure 7–19. Propionyl-CoA is converted to succinyl-CoA for further catabolism by the citric acid cycle.

Impairment of the reaction catalyzed by methylmalonyl-CoA mutase leads to the accumulation of methylmalonate, normally a minor product of the pathway. This may be caused by a genetic deficiency of the enzyme, by a dietary deficiency of vitamin B_{12}, or by a block in the absorption or further metabolism of the vitamin. Absorption of vitamin B_{12} is mediated by **intrinsic factor,** a vitamin B_{12}-binding protein that is secreted into the stomach. Genetic deficiency of intrinsic factor, **pernicious anemia,** results in a disorder characterized by symptoms of vitamin B_{12} deficiency. (See Chapter 12 for further discussion of vitamin B_{12} deficiency.)

Ketone Bodies

* An important function of the liver is to produce 2 compounds—acetoacetate and β-hydroxybutyrate—that can be used as fuels by nonhepatic tissues (Fig 7–20).* These 4-carbon compounds are manufactured by the mitochondria of the liver and are consumed by the brain and muscle.* For the brain, which is unable to take up fatty acids, these compounds represent an important alternative to glucose. β-Hydroxybutyrate is formed in liver by the NADH-dependent reduction of acetoacetate. Thus, the ratio of β-hydroxybutyrate to acetoacetate in the circulation reflects the ratio of NADH to NAD in hepatic mitochondria. A portion of the acetoacetate in circulation decomposes, forming CO_2 and acetone. Collectively, acetoacetate, β-hydroxybutyrate, and acetone are termed ketone bodies.

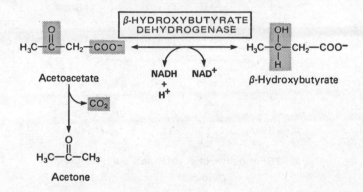

Figure 7–20. Interrelationships of the ketone bodies.

The key intermediate in ketone body production (**ketogenesis**) is acetoacetyl-CoA. This compound can be formed both as the result of the incomplete breakdown of fatty acids in β-oxidation and through the condensation of 2 acetyl-CoA units.* Because both fatty acids and amino acids contribute to the mitochondrial acetyl-CoA pool, both can contribute substrates for ketogenesis. The contribution from fatty acids is quantitatively more important than that from amino acids. Once formed, acetoacetyl-CoA can be converted to acetoacetate either by simple deacylation or by a more complex route. In the latter, which is quantitatively more important, acetoacetyl-CoA and acetyl-CoA condense to form β-hydroxy-β-methylglutaryl-CoA (HMG-CoA), which is then cleaved to yield acetoacetate and acetyl-CoA (Fig 7–21).

In tissues that consume ketone bodies, primarily brain and muscle, β-hydroxybutyrate is oxidized to form acetoacetate, and the latter is activated for further metabolism by attachment to CoA. Activation is accomplished

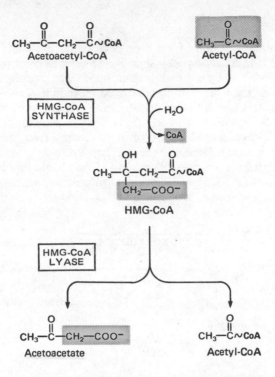

Figure 7-21. Formation of acetoacetate via HMG-CoA.

via an exchange reaction in which acetoacetate replaces the succinyl group of succinyl-CoA.

Citric Acid Cycle

◦The enzymes of the citric acid cycle (also called the **tricarboxylic acid [TCA] cycle** or the **Krebs cycle**) provide the means to oxidize the acetyl-CoA generated by the partial catabolism of carbohydrates and lipids. These enzymes are located within the matrix and on the inner membrane of the mitochondria.

The reactions of the citric acid cycle are illustrated in Fig 7-22. (1) Acetyl-CoA enters the cycle by condensing with oxaloacetate to form citrate. The energy needed to drive this reaction is provided by the high-energy thioester bond of acetyl-CoA. (2) Citrate is converted to isocitrate, which (3) is oxidatively decarboxylated to form α-ketoglutarate and CO_2, while NAD is reduced to NADH. (4) α-Ketoglutarate is also oxidatively decarboxylated. The products of the reaction are succinyl-CoA, CO_2, and NADH. The enzyme that catalyzes reaction 4, α-ketoglutarate dehydrogenase, is a mul-

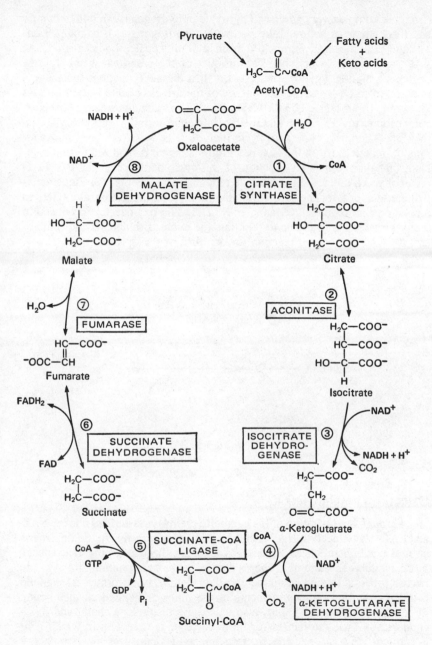

Figure 7–22. The citric acid cycle.

tienzyme complex very similar to pyruvate dehydrogenase in both structure and mechanism of action. Like pyruvate dehydrogenase, it contains 3 enzymatic activities and requires thiamin pyrophosphate, lipoic acid, CoA, FAD, and NAD for activity. The product of the reaction, succinyl-CoA, contains a high-energy thioester bond. (5) In the next reaction, succinate is released from CoA and the free energy of the thioester bond is used to form guanosine triphosphate (GTP). GTP, like ATP, is a nucleoside triphosphate; it contributes to ATP production by transferring its terminal phosphoryl group to ADP. Thus, the citric acid cycle includes one substrate level phosphorylation reaction. (6) In the next reaction, succinate is oxidized to fumarate. Note that this is the only reaction of the citric acid cycle that uses FAD rather than NAD as the acceptor of reducing equivalents. (7) Fumarate is hydrated to form malate, and (8) malate is then oxidized to oxaloacetate in a reaction that produces yet another NADH. The oxaloacetate produced in the last reaction is available to reinitiate the cycle with another molecule of acetyl-CoA. The overall reaction for one turn of the cycle is

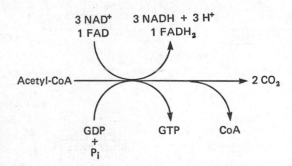

Oxidative Phosphorylation

Through the operation of the catabolic pathways described above, NAD and FAD are reduced to NADH and $FADH_2$, respectively. In the final stage of fuel catabolism, these coenzymes are reoxidized, and the free energy made available through this process drives the phosphorylation of ADP. The overall process, termed oxidative phosphorylation, is performed through the coupled operation of 2 components of the inner mitochondrial membrane: the **electron transport chain,** also called the respiratory chain; and mitochondrial ATPase (so named for the reverse of the reaction in which ATP is formed) (Fig 7–23). The electron transport chain uses the free energy available from reoxidation of NADH and $FADH_2$ to create a transmembrane proton gradient. Mitochondrial ATPase uses the energy stored in this gradient to form ATP from ADP and P_i.

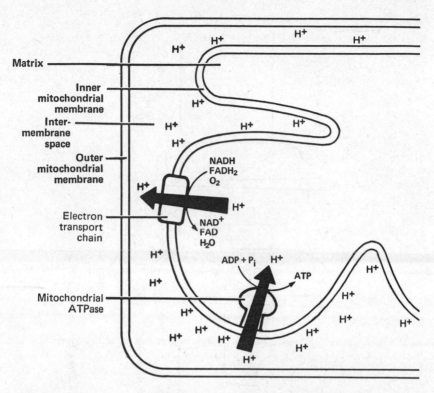

Figure 7–23. Oxidative phosphorylation is performed by the coupled activities of the electron transport chain and mitochondrial ATPase.

The electron transport chain consists of a series of electron carriers that are associated with the inner mitochondrial membrane (Fig 7–24). The chain accepts electrons from NADH and $FADH_2$ and ultimately donates them to oxygen, forming water. Most components of the chain are proteins whose prosthetic groups can undergo oxidation and reduction. The first protein, NADH dehydrogenase, employs 2 types of prosthetic group: flavin mononucleotide (FMN), which like FAD is derived from riboflavin, and clusters of iron and sulfur atoms, known as Fe-S centers. The cytochromes use both heme and Fe-S centers as electron carriers. Coenzyme Q (also called ubiquinone) is a low-molecular-weight lipid that moves within the mitochondrial membrane, shuttling electrons between different components of the electron transport chain (Fig 7–25).

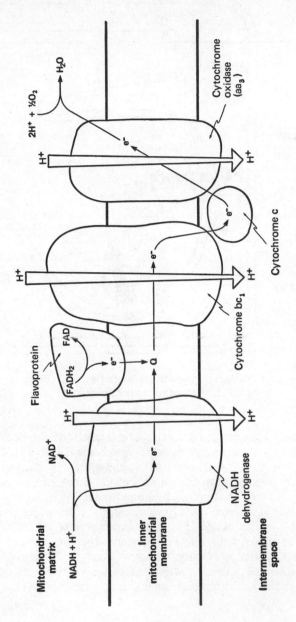

Figure 7–24. The electron transport chain.

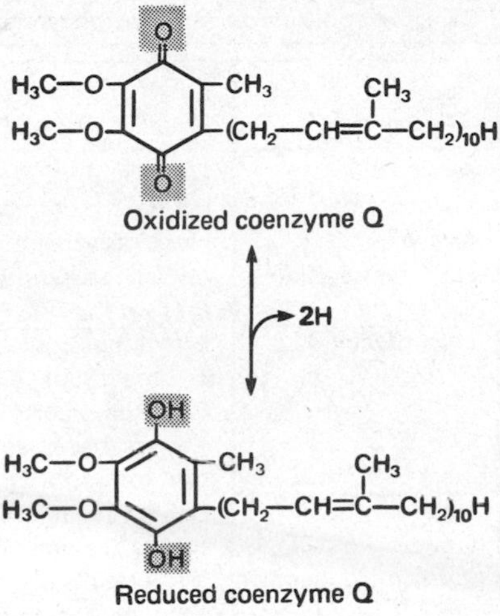

Oxidized coenzyme Q

$\rightarrow$ 2H

Reduced coenzyme Q

Figure 7–25. Coenzyme Q, the mobile electron carrier of the electron transport chain.

Reducing equivalents enter the electron transport chain at 2 different sites. NADH donates its electrons to NADH dehydrogenase. They are passed successively to coenzyme Q, the cytochromes, and finally O_2. $FADH_2$ is not so strong a reducing agent as NADH and is unable to reduce the electron carriers of NADH dehydrogenase. Instead, it donates electrons to coenzyme Q, and from there they pass along the chain to O_2.

As electrons are transported along the chain, protons are pumped from the mitochondrial matrix to the space between the inner and outer mitochondrial membranes. The mechanism by which this happens is still unknown. It is clear, however, that there are 3 proton-pumping components of the electron transport chain: the NADH dehydrogenase complex, the cytochrome bc_1 complex, and the cytochrome aa_3 complex (cytochrome oxidase). Each of these pumps extrudes protons from the mitochondrion each time a pair of electrons is transferred through it. Because the electrons of NADH enter the chain at NADH dehydrogenase, they activate all 3 pumps. Those of $FADH_2$, entering the chain at coenzyme Q, activate only 2.

The activity of the pumps establishes a proton gradient spanning the inner mitochondrial membrane. Protons reenter the mitochondrion through mitochondrial ATPase, driving the formation of ATP by an unknown mechanism. In the coupled process of electron transport and ATP synthesis, **oxidation of 1 NADH results in the production of 3 ATP and oxidation of 1 FADH₂ results in 2 ATP.**

Although electron transport and ATP synthesis are performed by physically distinct structures within the mitochondrial membrane, they are functionally **coupled** via the transmembrane proton gradient. This coupling is

Table 7—1. Effects of poisons on oxidative phosphorylation.

Type of Poison	Examples	Consequences for Electron Transport and ATP Synthesis
Electron transport inhibitor	Antimycin A Carbon monoxide Cyanide Hydrogen sulfide Rotenone	Direct inhibition of electron transport results in a *decrease* in the transmembrane proton gradient. Without this driving force, mitochondrial ATP synthesis stops.
Mitochondrial ATPase inhibitor	Oligomycin	Direct inhibition of mitochondrial ATPase results in an *increase* in the transmembrane proton gradient. As the gradient becomes larger, it inhibits electron transport.
Uncoupling agent	Valinomycin Gramicidin 2,4-Dinitrophenol	An increase in the permeability to protons of the inner mitochondrial membrane results in a *decrease* in the transmembrane gradient. As the gradient is dissipated, ATP synthesis declines (no driving force) *but* electron transport continues (no inhibition).

revealed by the effects of a number of compounds that inhibit oxidative phosphorylation (Table 7–1). Rotenone (a plant product used as an insecticide), antimycin A (an antibiotic), cyanide, carbon monoxide, and hydrogen sulfide all inhibit components of the electron transport chain. When electron transport is blocked, no protons are pumped; and because there is no proton gradient, ATP synthesis stops. Oligomycin (an antibiotic) directly inhibits mitochondrial ATPase. When this enzyme is inhibited, electron transport also stops. This result suggests that there is a limit to the magnitude of the transmembrane proton gradient and that electron transport is blocked when the gradient reaches its upper limit.

Although oxidative phosphorylation and electron transport are normally functionally coupled, they can be **uncoupled** by a number of compounds that make the mitochondrial membrane permeable to protons. Valinomycin, gramicidin, and 2,4-dinitrophenol all allow protons to reenter the mitochondrial matrix without passing through mitochondrial ATPase. When mitochondria are poisoned by these substances, the proton gradient is rapidly dissipated and ATP synthesis stops. Furthermore, in the presence of uncoupling agents, the normal control imposed on electron transport by the proton gradient is lost. Under these conditions, electron transport proceeds as fast as is permitted by the availability of NADH, $FADH_2$, and O_2. The free energy that normally is used to form ATP is lost as heat.

ATP Yield & O_2 Consumption

Knowing all of the reactions of fue[...]
both the amount of ATP formed and the am[...]
of each fuel (Table 7–2).*Glycolysis can op[...]
aerobically.*For each glucose consumed aerobi[...]
the malate shuttle is used and 36 if the glycerol p[...]
*Oxidation of 1 glucose requires 6 O_2. Thus, in the [...]
6–6.3 ATP are produced for every O_2 consumed.*Ope[...]
glycolysis yields only 2 ATP per glucose. Clearly, it is hig[...]
to fully oxidize glucose whenever possible.

In fatty acid catabolism, the ATP yield and oxygen consump[...]
upon the chain length of the fatty acid. Table 7–2 presents the to[...]
for palmitate. It is evident that the consequences of combustion of [...]
hydrates and fats are different.*Fats yield a larger amount of ATP (8 A[...]
carbon) than do carbohydrates but require a larger consumption of oxygen[...]

Regulation of Catabolism

It is of course important to regulate the catabolism of fuels so that they
are not wasted. The pathways of catabolism are elegantly regulated by a
combination of controls that include allosteric regulation, covalent modification of enzymes, and availability of substrates. The system of regulation
is designed to produce the following results:

(1) A nearly constant supply of ATP is maintained in every cell, and
only as much fuel is consumed as is needed to meet the demands for ATP
production.

(2) When the demand for ATP exceeds the maximum rate of aerobic
metabolism (that rate being limited by the rate of delivery of O_2 to the
tissues), the activity of glycolysis is increased relative to that of the strictly
aerobic pathways of catabolism, β-oxidation, the citric acid cycle, and
oxidative phosphorylation.

(3) In tissues that can use both fatty acids and glucose, the catabolism
of glucose is reduced when both fats and oxygen are available.

These results are produced by the regulatory devices summarized in Fig
7–26.

The rate at which **mitochondrial ATPase** catalyzes the formation of
ATP is limited by the availability of its substrate ADP. When ATP stores
are full, very little ADP is available and mitochondrial ATPase activity is
reduced. This event in turn regulates electron transport. When ATP synthesis
stops, electrons have no means of reentering the mitochondrion. Because
there is a limit to the size of the proton gradient spanning the inner mitochondrial membrane, the halt in ATP synthesis also inhibits the operation
of electron transport, leading to an accumulation of NADH and $FADH_2$ in
the mitochondria.

l catabolism, we can now calculate
ount of O_2 used in the catabolism
erate both aerobically and an-
cally, 38 ATP are formed if
phosphate shuttle is used.
catabolism of glucose,
ting anaerobically,
ly advantageous

tion depend
al yields
carbo-
TP/

Table 7–2. Summary of ATP yield and oxygen cons[umption]

Fuel and Pathway	Direct Products of Pathway	ATP Yield After Oxidative Phosphorylation		
Anaerobic glycolysis (glucose → 2 lactate)	2 ATP			
Complete oxidation of glucose				
Glycolysis (glucose → 2 pyruvate)	2 ATP, 2 NADH	6–8 ATP*		
Pyruvate dehydrogenase (2 pyruvate → 2 acetyl-CoA)	2 NADH	6 ATP		
Citric acid cycle (2)(2 acetyl-CoA → 4 CO_2)	6 NADH, 2 FADH$_2$, 2 GTP	24 ATP		
		36–38 ATP*		
Complete oxidation of palmitate				
β-oxidation (palmitate → 8 acetyl-CoA)	–2 ATP, 7 NADH, 7 FADH$_2$	33 ATP	16 O_2	8.1
Citric acid cycle (8 acetyl-CoA → 16 CO_2)	24 NADH, 8 FADH$_2$, 8 GTP	96 ATP		
		129 ATP	23 O_2	5.6

*The higher yield of ATP is obtained when the malate shuttle is used.

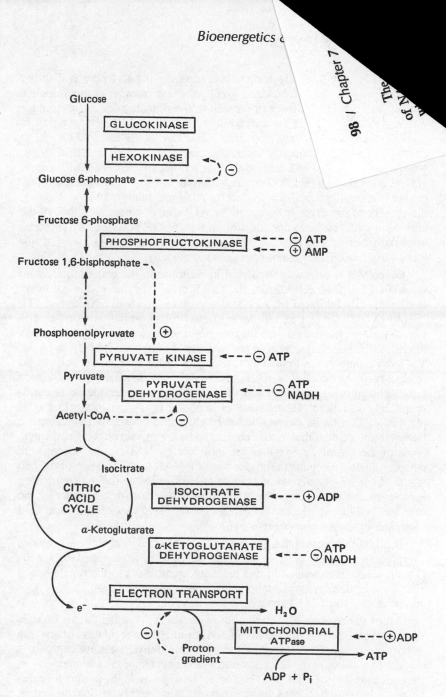

Figure 7–26. Regulation of ATP production.

activity of the citric acid cycle is regulated in part by the availability AD and FAD. Mitochondrial levels of these electron carriers decline henever the rate of delivery of oxygen to the mitochondria is limiting for electron transport. Under these conditions, NADH and $FADH_2$ accumulate at the expense of their oxidized forms. Because NAD and FAD serve as cosubstrates for the oxidation-reduction reactions of the citric acid cycle, when their concentrations decline these reactions are slowed. Furthermore, NADH allosterically inhibits **isocitrate dehydrogenase.** Isocitrate dehydrogenase is also allosterically activated by ADP and inhibited by ATP. Thus, the activity of the cycle is regulated by ATP usage. Note that although the citric acid cycle includes one reaction in which ATP is formed by substrate level phosphorylation, because the cycle is inhibited by high levels of mitochondrial NADH, it cannot operate anaerobically.

Glycolysis is primarily regulated in response to the cell's requirement for ATP. **Phosphofructokinase,** the rate-limiting enzyme of the pathway, is allosterically inhibited by ATP and stimulated by AMP. Note that ATP is both a substrate and an allosteric effector of phosphofructokinase. **Pyruvate kinase,** which catalyzes the rate-limiting reaction in the second phase of glycolysis, is inhibited by ATP and activated by fructose 1,6-bisphosphate. The latter control helps to match the activities of the first and second phases of glycolysis. As the activity of phosphofructokinase increases, more fructose 1,6-bisphosphate is produced and the activity of pyruvate kinase is correspondingly increased. **Hexokinase** is inhibited by its product, glucose 6-phosphate. This limits the production of glucose 6-phosphate in nonhepatic tissues to the amount that can be consumed by glycolysis or other pathways. None of the glycolytic enzymes are inhibited by NADH, and because the cytoplasm has a mechanism for regenerating NAD in the absence of oxygen (the Cori cycle), glycolysis can operate anaerobically. In rapidly contracting muscle, as the requirement for ATP exceeds the oxidative capacity of the cell, the activity of glycolysis increases relative to that of the citric acid cycle and oxidative phosphorylation.

In addition to contributing to the production of ATP, glycolysis provides intermediates that are used in the synthesis of fatty acids (see Chapter 9). Citrate, a key intermediate in this pathway, contributes to the regulation of glycolysis by inhibiting phosphofructokinase. Hormonal regulation of glycolysis is discussed in Chapter 8.

All of the controls discussed above operate through limitations on substrate availability, end product inhibition, or allosteric regulation of enzyme activity. In contrast, regulation of the **pyruvate dehydrogenase** complex is mediated by covalent modification of the first enzyme of the complex. When the concentration of ATP, NADH, or acetyl-CoA is high, pyruvate decarboxylase is phosphorylated by an enzyme called pyruvate dehydrogenase kinase. Phosphorylation decreases the activity of the complex. The regulatory phosphate molecule is removed by another enzyme, pyruvate dehydrogenase

phosphatase, which is active at a low rate at all times. When
and acetyl-CoA levels decline, the rate of phosphate remov;
rate of phosphate addition and the enzyme is reactivated. The
of this regulation is to inhibit pyruvate dehydrogenase when ₍
filled, when NADH accumulates and is not being reoxidized by the electron
transport chain (an indication of either sufficient ATP or limiting O_2), or
when acetyl-CoA is being produced through the metabolism of other fuels.

Together these regulatory devices serve to provide a pool of ATP that
meets the energy needs of the cell while conserving the supply of fuel
available to that cell. In Chapter 8, we see how additional controls on these
basic pathways coordinate fuel usage throughout the body.

8 | Glucose Storage & Homeostasis

OBJECTIVES

● Know the normal range of blood glucose concentration. Be able to list the metabolic processes involved in maintaining blood glucose within the normal range. Know when in the cycle of feeding and fasting each of these processes operates.

● Know the intermediates and enzymes involved in glycogenesis and glycogenolysis. Be able to trace the pathways by which glucose, fructose, and galactose are incorporated into glycogen.

● Be able to explain how glycogen synthesis and breakdown are regulated and to summarize the differences in the systems that regulate glycogen metabolism in liver and muscle.

● Be able to name and describe the reactions that distinguish gluconeogenesis from glycolysis and to describe the roles of fatty acids and amino acids in this process.

● Be able to explain how the activities of glycolysis and gluconeogenesis are regulated by the nutritional status of the body.

THE HUMAN BODY has a continuous need for ATP, and yet the fuels used in ATP production are taken into the body discontinuously in meals. During digestion, enough nutrients are absorbed to meet the energy demands of the body until the next meal. However, the absorbed nutrients cannot remain in circulation until needed. To overcome this problem, the body converts excess nutrients taken in during feeding to storage forms of fuel and retrieves those fuels from storage during fasting. Fuel is stored in 3 forms: glycogen (a polymer of glucose), triglycerides (each containing 3 fatty acids esterified to a molecule of glycerol), and proteins. During fasting, each of these compounds is catabolized to meet the demand for fuel. The sole function of glycogen and triglycerides in human metabolism is that of fuel·storage. In contrast, proteins are formed primarily to act as catalysts, carriers, receptors, and structural components of the body. Thus, consumption of protein as a fuel during starvation compromises important metabolic functions.

Hormonal Control of Fuel Metabolism

A **hormone** is a chemical signal synthesized by one type of tissue and carried by the circulatory system to another tissue, the target tissue, where it elicits a specific biochemical response. A number of hormones play a role in the regulation of fuel storage. This chapter discusses only the effects of the pancreatic hormones insulin and glucagon on fuel metabolism.

During digestion, glucose and amino acids absorbed from the intestine enter the circulation and are distributed throughout the body. The pancreas monitors the concentrations of those nutrients in circulation and, through the secretion of insulin and glucagon, regulates fuel usage by other tissues of the body. **Insulin,** an anabolic hormone, stimulates the synthesis of the macromolecular components of cells and leads to the storage of excess fuels. **Glucagon,** a catabolic hormone, limits the synthesis of macromolecules and causes the release of stored nutrients. An increase in the concentration of glucose in circulation leads to an increase in the secretion of insulin and a decrease in the secretion of glucagon. Conversely, a decrease in blood glucose leads to a decrease in the secretion of insulin and an increase in the secretion of glucagon. Insulin is largely responsible for maintaining the inverse relationship between the concentrations of insulin and glucagon by inhibiting the secretion of glucagon from the pancreas.

Normal Blood Glucose Levels

An important factor in the cycle of fuel storage and consumption of stores is the strict requirement of certain tissues for glucose. We have already seen that mature red blood cells lack mitochondria and therefore can utilize only glucose as a fuel. Under normal conditions, the brain is also entirely dependent upon a continuous supply of glucose because it is unable to take up fatty acids and can absorb ketone bodies only after several days of fasting when they reach a high concentration in the blood. To meet the fuel requirements of tissues dependent on glucose, the blood glucose concentration is maintained within the limits of 3–7 mmol/L. This is in striking contrast to the wide variations in concentration of fatty acids (10-fold) and ketone bodies (100-fold). Maintenance of blood glucose within the normal range is termed **glucose homeostasis.**

There are several important reasons why the blood glucose concentration is narrowly regulated. If the blood glucose concentration falls below 1.5 mmol/L, the brain is inadequately supplied with fuel, brain ATP levels begin to drop, and brain function is impaired. Coma or death may ensue. The safe upper limit on blood glucose is set by its osmotic and chemical properties. High blood glucose levels have the effect of dehydrating tissues, and as water moves into circulation, the tissues also lose important ions. High blood glucose levels also accelerate one of the processes of protein aging, nonenzymatic glycosylation. In its open-chain form, glucose reacts nonenzymatically with exposed amino groups of proteins and, in so doing, changes the catalytic and structural properties of the proteins. Hemoglobin, collagen, proteins of the lens of the eye, and a number of other proteins all suffer glucose modification in proportion to the concentration of glucose in circulation.

The concentration of glucose in circulation is controlled by stimulating its use and diverting it into stores when it is abundant and by both limiting its use and actively synthesizing it when it is scarce.

During digestion of a meal, blood glucose levels rise. The accompanying increase in circulating insulin levels leads to the removal of glucose from circulation by increasing the rate at which glucose is transported into cells and by increasing the activities of glucose-consuming pathways. Glucose enters cells efficiently only when transported by a specific carrier protein located in the surface (plasma) membrane of the cell. Glucose entry into some tissues, eg, brain, red blood cells, and liver, is not regulated and depends solely on the concentration of glucose in the circulation. The glucose transport systems of other tissues, eg, adipose tissue and muscle, require insulin for activity. Thus, tissues that have insulin-regulated glucose carriers take up glucose only when glucose is abundant.

In addition to consuming glucose for ATP production, most cells convert excess glucose to storage molecules. The liver, muscle, and other tissues polymerize glucose to form glycogen (**glycogenesis**), and the liver and adipose tissue convert glucose to fatty acids (**lipogenesis**), which are then stored by adipose tissue as triglycerides.

When glucose is not being absorbed from the digestive tract, blood glucose and insulin levels begin to decline. At that time, the use of glucose is radically curtailed, and only those tissues that are dependent on glucose are able to absorb it. Tissues that are denied glucose catabolize fats instead. To maintain the blood glucose levels as fasting progresses, the liver breaks down glycogen to glucose (**glycogenolysis**) and synthesizes glucose de novo (**gluconeogenesis**) using amino acids as substrates.

Fig 8–1 illustrates how blood glucose levels are maintained at various times after a meal. In the initial several hours, glucose is absorbed from the intestine. When this supply has been depleted, first glycogenolysis and then gluconeogenesis contribute glucose to the blood.

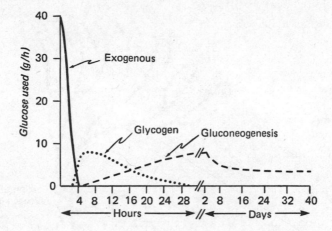

Figure 8–1. The sources of blood glucose in glucose homeostasis. (Redrawn and reproduced, with permission, from Hanson RW, Mehlman MA [editors]: *Gluconeogenesis: Its Regulation in Mammalian Species.* Wiley, 1976.)

Glucokinase

Liver cells contain 2 enzymes that catalyze the phosphorylation of glucose: hexokinase and glucokinase. The K_m of glucokinase for glucose is 20 mmol/L—considerably above the normal blood glucose levels. Thus, in normal individuals, glucokinase is not saturated with glucose and its activity fluctuates as blood glucose concentrations change. As the concentration of glucose in circulation increases, glucokinase helps to maintain normal blood glucose levels by phosphorylating an increasing fraction of the glucose that passes through the liver. Following a meal, the concentration of glucose in the blood entering the liver may be as high as 15 mmol/L, while the blood leaving the liver contains only 7 mmol/L of glucose. Glucokinase is responsible for the difference. The glucose 6-phosphate formed by glucokinase is both stored as glycogen and catabolized to form substrates for fatty acid synthesis.

Glycogen Metabolism

Glycogen is a highly branched polymer of glucose found in the cytoplasm of most cells in the form of large granules, 10–40 nm in diameter (Fig 8–2). Glycogen consists of chains of glucose units in $\alpha(1\rightarrow4)$-glycosidic linkages with $\alpha(1\rightarrow6)$-glycosidic bonds forming the branch points. The enzymes that catalyze glycogen synthesis and breakdown are bound to the surface of the granule. Glycogen is an important form of stored fuel because, unlike fat, it can be converted to glucose and so help maintain blood glucose levels.

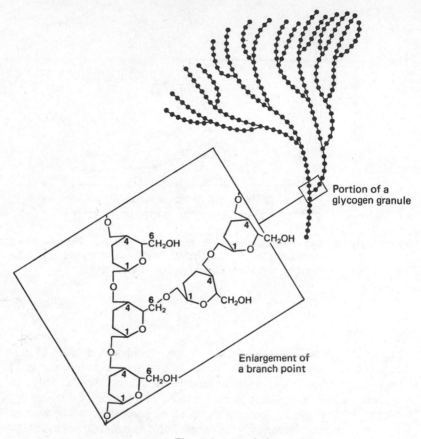

Portion of a
glycogen granule

Enlargement of
a branch point

Figure 8-2. The structure of glycogen.

Although most tissues contain some glycogen, the major stores are in liver and muscle. The glycogen of liver is primarily used to supply glucose to other tissues during fasting. After 12–18 hours of fasting, liver glycogen is almost totally depleted. Muscle glycogen is used within the cell in which it is stored as a fuel for muscle contraction. Its stores are depleted mainly by muscle use.

The synthesis of glycogen is an energy-requiring process. Indirectly, the energy is provided by ATP, but the immediate source is uridine triphosphate (UTP). UTP is used to synthesize an "activated" form of glucose, uridine diphosphate glucose (UDP-glucose), which is the substrate for glycogen synthesis. Recall that UDP-glucose is also produced through the metabolism of galactose (see p 76). Formation of UDP-glucose starts with the ATP-dependent phosphorylation of glucose by glucokinase in liver and hexokinase in muscle (Fig 8–3). Glucose 6-phosphate is isomerized to glu-

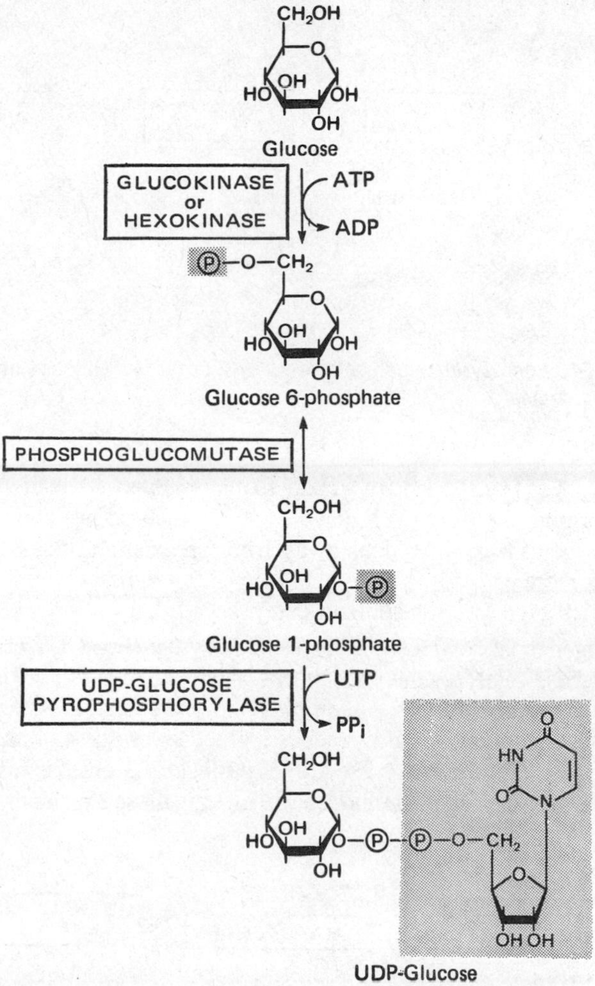

Figure 8–3. The first step in glycogen synthesis is formation of UDP-glucose. (P) = phosphoryl group.

cose 1-phosphate by phosphoglucomutase. Formation of UDP-glucose from glucose 1-phosphate and UTP is catalyzed by UDP-glucose pyrophosphorylase.

Incorporation of the glucose moiety of UDP-glucose into glycogen involves 2 additional enzymes (Fig 8–4). (1) **Glycogen synthase** catalyzes the addition of glucose units to the ends of preexisting polymers that are at least 4 units in length, forming the $\alpha(1\rightarrow4)$ linkages of the polymer. (2) New branches are formed by a **branching enzyme** that transfers a unit of 6 or 7

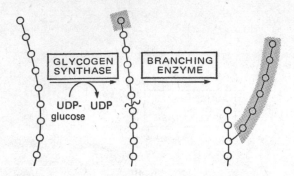

Figure 8–4. Glycogen synthesis is catalyzed by 2 enzymes. Glucose units are represented by circles.

glucose residues from the end of a chain to a more interior site in the polymer, thereby forming the α(1→6) linkage of the branch point. The new branch point must be at least 4 residues away from preexisting branches. Because it provides more ends for synthesis and breakdown, branching increases the rate at which glycogen metabolism can proceed.

Glycogen breakdown is catalyzed by 3 enzymes (Fig 8–5). (1) **Glycogen phosphorylase** catalyzes the phosphorolysis of the α(1→4) linkages, yielding glucose 1-phosphate. Glycogen phosphorylase acts on the ends of glycogen chains, stopping when it reaches a glucose residue 4 units away from a branch point. (2) A **transferase** moves a block of 3 glucose residues from a branch point to the end of another chain, making those 3 residues susceptible

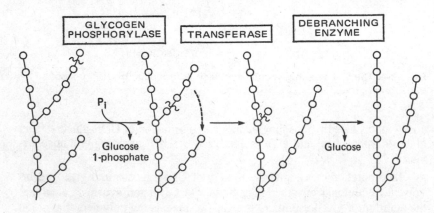

Figure 8–5. Glycogen degradation is catalyzed by 3 enzymes. Glucose units are represented by circles.

to glycogen phosphorylase. (3) A debranching en:
dase, releases the single glucose remaining at the l

Note that with the exception of those residues re
ing enzyme, the product of glycogen breakdown
This compound is converted by phosphoglucomutas
The glucose units stored in liver are destined to be
Because phosphorylated sugars cannot traverse the cell membrane, glucose
6-phosphate must be dephosphorylated. Liver, but not muscle, contains the
enzyme that carries out the final reaction in glucose production, glucose 6-
phosphatase. Muscle glycogen, on the other hand, serves as a store of glucose
to be used within the muscle cell during contraction. In muscle, the glucose
1-phosphate released in glycogenolysis is isomerized to glucose 6-phosphate,
which directly enters glycolysis.

Regulation of Glycogen Metabolism

We have seen that glycogen synthesis and breakdown are performed by
2 distinct groups of enzymes. The use of separate enzymes for anabolic and
catabolic pathways is not unique to glycogen metabolism but rather is a
general principle of biology. By using separate pathways, synthesis and
breakdown can both be made energetically favorable. However, one consequence of this design is that together the pathways potentially form an
energy-consuming cycle (Fig 8–6). In order to prevent waste of UTP, synthesis and breakdown of glycogen must be coordinated. In addition, glycogen

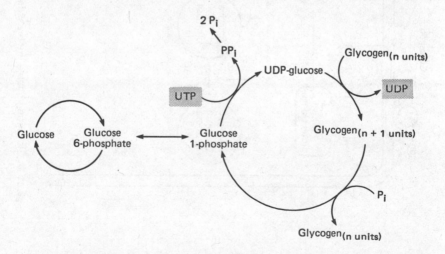

Figure 8–6. The pathways of glycogen synthesis and breakdown form a cycle.

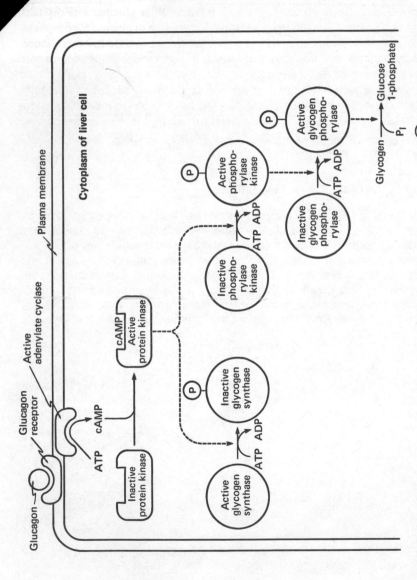

Figure 8–7. Glucagon initiates the breakdown of liver glycogen. – – – → = Catalysis. Ⓟ = phosphoryl group.

metabolism must be made responsive to the nutritional status of the body and tailored to the roles of the individual tissues.

Glycogen metabolism in the liver is regulated primarily by the hormones insulin and glucagon and by blood glucose levels. When blood glucose levels are low, secretion of glucagon from the pancreas increases. Glucagon triggers glycogen breakdown in the liver via the cascade of reactions shown in Fig 8–7. Glucagon interacts with the target liver cells by binding to a **receptor** protein located on the outside surface of the plasma membrane. When the receptor is occupied, it interacts with and activates adenylate cyclase, an enzyme that is embedded in the plasma membrane. Adenylate cyclase catalyzes the conversion of ATP to cyclic adenosine monophosphate (cAMP) (Fig 8–8). cAMP mediates the effects of glucagon within the liver cell and therefore is referred to as a **second messenger** of glucagon. cAMP also acts as the second messenger for a number of other hormones.

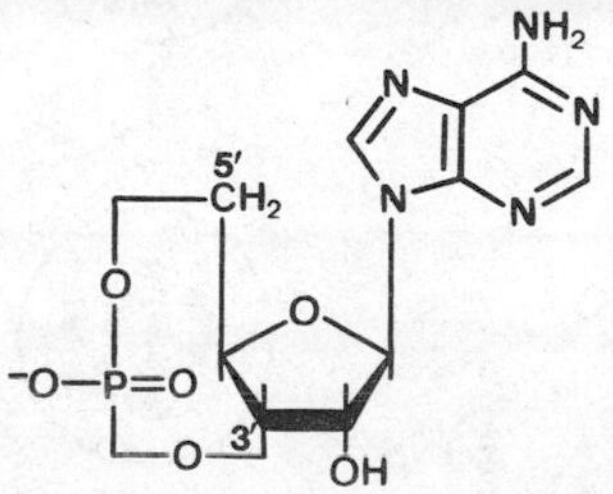

Figure 8–8. The structure of cAMP.

cAMP acts as the allosteric activator of a cAMP-dependent protein kinase. Protein kinase in turn controls the activities of a number of enzymes that are regulated by **phosphorylation** and **dephosphorylation** (see p 54). Two of the substrates of the cAMP-dependent protein kinase are glycogen synthase and phosphorylase kinase. While protein kinase phosphorylates both proteins, phosphorylation has opposite effects on their activities— phosphorylation *inactivates* glycogen synthase and *activates* phosphorylase kinase. The function of phosphorylase kinase is to phosphorylate and thereby activate glycogen phosphorylase. Thus, through the formation of cAMP and protein phosphorylation, glucagon has the effect of coordinately turning on glycogen breakdown and turning off glycogen synthesis. (By convention, the phosphorylated form of each of these enzymes is referred to as the "a" form, eg, glycogen phosphorylase a, and the dephosphorylated form as the "b" form.)

The activity of hepatic glycogen phosphorylase is also regulated by a number of allosteric effectors. The key effector for glucose homeostasis is glucose. When glucose is abundant, it binds to glycogen phosphorylase and

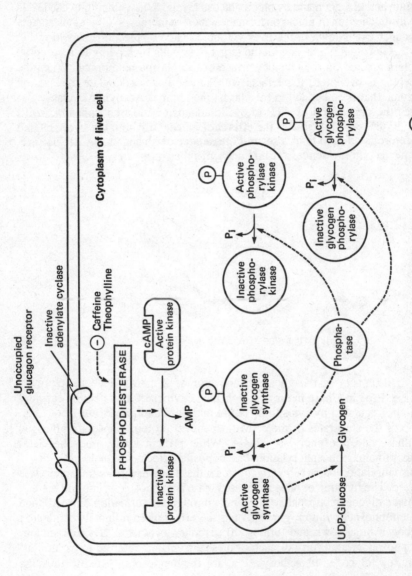

Figure 8–9. Withdrawal of glucagon reverses the regulatory cascade. – – – → = Catalysis. Ⓟ = phosphoryl group.

stabilizes an inactive conformation of the enzyme. Thus, as the concentration of glucose in the liver rises, glycogen phosphorylase is inhibited and glycogen breakdown is stopped.

An increase in the concentration of glucose also turns off the regulatory cascade initiated by glucagon. Rising glucose levels stimulate the secretion of insulin, which in turn depresses the secretion of glucagon. Because each of the steps of the regulatory cascade is subject to an opposing reaction, when glucagon is withdrawn the opposing reactions take over (Fig 8–9). A phosphodiesterase degrades cAMP to AMP, which is not an activator of protein kinase, and a phosphatase removes the regulatory phosphate groups from the enzymes of the cascade. Caffeine and theophylline, found in coffee and tea, respectively, inhibit the phosphodiesterase that degrades cAMP and therefore prolong the effects of glucagon and other hormones that act via cAMP.

In addition to its indirect effects via regulation of glucagon levels, insulin also has a number of direct effects on fuel metabolism. At this time, we know relatively little about how the latter are mediated. We do know that insulin (like glucagon) acts by binding to a cell surface receptor; that the presence of insulin leads to the dephosphorylation of many enzymes regulated by phosphorylation and dephosphorylation; and that insulin induces the synthesis of some enzymes. As yet, however, the events between binding of the hormone to its receptor and its ultimate effects are unknown. Insulin has a major effect on glycogen metabolism by favoring the dephosphorylation of glycogen synthase, phosphorylase kinase, and glycogen phosphorylase, which in turn activates glycogen synthesis and inhibits glycogen breakdown. Insulin also regulates the synthesis of glucokinase and thereby increases the capacity of the liver to phosphorylate glucose as it enters the liver cell.

Glycogen metabolism of muscle cells is controlled by a regulatory cascade similar to that of liver. Skeletal muscle cells degrade glycogen during contraction to provide substrates for the glycolytic pathway in the muscle cell. As in the liver, muscle cells respond to insulin stimulation by synthesizing glycogen. However, in muscle, glycogenolysis is activated not by glucagon but by calcium, and epinephrine.

When a muscle cell is stimulated to contract, the concentration of calcium in the cell rises. The change in calcium concentration influences the activity of phosphorylase kinase, one subunit of which is a calcium-binding protein named **calmodulin**. Binding of calcium to calmodulin activates phosphorylase kinase, which in turn activates glycogen phosphorylase and inactivates glycogen synthase (Fig 8–10). Glycogenolysis can also be activated hormonally. Stress in many forms triggers the release of the catecholamine hormone **epinephrine** from the adrenals. Epinephrine stimulation prepares the muscles for imminent contraction. Like glucagon, epinephrine binds to its target cells via receptors (the adrenergic receptors) located in the plasma membrane. Binding of epinephrine to the adrenergic receptors on skeletal

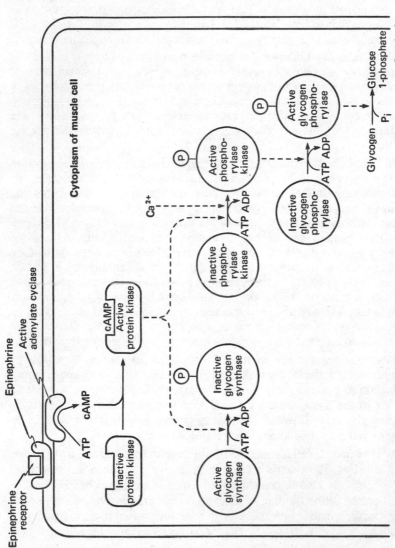

Figure 8–10. Epinephrine and Ca²⁺ stimulate glycogen breakdown in muscle cells. – – – ⇒ = Catalysis. Ⓟ = phosphoryl group.

ate. One NADH is consumed in the conversion of 1,3-bisphosphoglycerate to glyceraldehyde 3-phosphate. Because 2 mol of pyruvate are used in the synthesis of one glucose, each mole of glucose synthesized in gluconeogenesis costs the cell 6 ATP and 2 NADH. Glycolysis and gluconeogenesis cannot operate at the same time, and therefore the ATP and NADH consumed in gluconeogenesis must come from the oxidation of other fuels, primarily fatty acids.

Although fats provide most of the energy for gluconeogenesis, they contribute only a small fraction of the carbon atoms used as substrates. This is a consequence of the structure of the citric acid cycle. The most abundant fatty acids in humans—those with an even number of carbon atoms—are degraded by the enzymes of β-oxidation to acetyl-CoA. Acetyl-CoA contributes 2-carbon fragments to the citric acid cycle, but early in the cycle, 2 carbons are lost as CO_2. Thus, metabolism of acetyl-CoA does not lead to an increase in the amount of oxaloacetate available for gluconeogenesis. If oxaloacetate is withdrawn from the cycle and not replaced, the ATP-forming capacity of the cell will soon be compromised. The citric acid cycle is not impaired during gluconeogenesis, because oxaloacetate is formed from pyruvate via the pyruvate carboxylase reaction.

Most of the carbon atoms consumed in glucose synthesis are provided by the catabolism of amino acids (Fig 8–12). Several of the common amino acids are degraded to pyruvate and therefore enter gluconeogenesis via the pyruvate carboxylase reaction. Others are converted to 4- and 5-carbon intermediates of the citric acid cycle and therefore can contribute to an increase in the oxaloacetate and malate content of the mitochondrion. Of

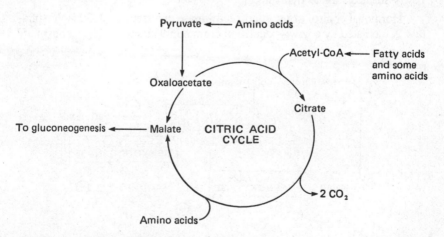

Figure 8–12. Carbon atoms used in the synthesis of glucose are derived from amino acids.

the 20 amino acids commonly found in proteins, only leucine and lysine are degraded entirely to acetyl-CoA and therefore cannot provide substrates for gluconeogenesis.

Regulation of Gluconeogenesis

Because the liver can both make glucose via gluconeogenesis and consume glucose via glycolysis, it must have a system of regulation that prevents the simultaneous operation of the 2 pathways. The regulatory system must also ensure that the metabolic activity of the liver is appropriate to the nutritional status of the body, ie, glucose production during fasting and glucose consumption in times of glucose abundance. The activities of gluconeogenesis and glycolysis are coordinately regulated by means of changes in the relative amounts of glucagon and insulin in circulation.

As blood glucose and insulin levels fall, fatty acids are mobilized from adipose stores and the activity of β-oxidation in liver increases. This results in an increase in the hepatic concentrations of both fatty acids and acetyl-CoA. Because amino acids are simultaneously mobilized from muscle, there is also an elevation of amino acid levels, primarily that of alanine. Hepatic amino acids are converted to pyruvate and other substrates of gluconeogenesis. The elevated levels of fatty acids, alanine, and acetyl-CoA all play a role in directing substrates into gluconeogenesis and preventing their consumption by the citric acid cycle (Fig 8–13). Acetyl-CoA allosterically activates pyruvate carboxylase and inhibits pyruvate dehydrogenase, thereby ensuring that pyruvate will be converted to oxaloacetate. Pyruvate kinase is inhibited by fatty acids and alanine, thus preventing the breakdown of the newly formed PEP to pyruvate.

Hormonal control of phosphofructokinase and fructose-1,6-bisphosphatase is mediated by a newly identified compound, fructose 2,6-bisphosphate

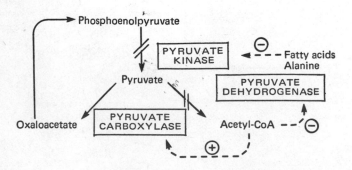

Figure 8–13. During fasting, fatty acids, alanine, and acetyl-CoA determine the fate of pyruvate in the liver.

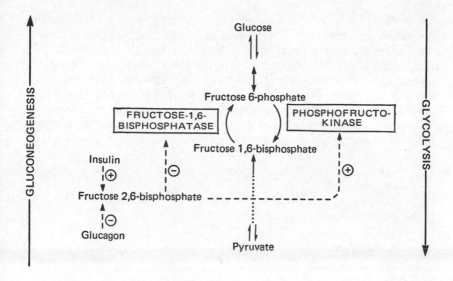

Figure 8–14. Fructose 2,6-bisphosphate controls the activities of phosphofructokinase and fructose-1,6-bisphosphatase.

(Fig 8–14). Formation and breakdown of this regulatory compound are catalyzed by enzymes that are controlled by phosphorylation and dephosphorylation. Changes in the concentration of fructose 2,6-bisphosphate parallel those for glucose and insulin; ie, its concentration increases when glucose is abundant and declines when glucose is scarce. Fructose 2,6-bisphosphate allosterically activates phosphofructokinase and inhibits fructose-1,6-bisphosphatase. Thus, when glucose is abundant, glycolysis is active and gluconeogenesis is inhibited. As glucose levels fall, the increase in glucagon results in a decrease in the hepatic concentration of fructose 2,6-bisphosphate and the coordinate inhibition of glycolysis and activation of gluconeogenesis.

Insulin Deficiency

Diabetes mellitus is a disease characterized by derangements of carbohydrate, lipid, and protein metabolism. Patients with type I (insulin-dependent) diabetes mellitus secrete little or no insulin. These individuals have abnormally elevated blood glucose levels if they do not receive insulin replacement therapy. Absence of insulin elevates blood glucose levels for 2 reasons: (1) the rate at which glucose is removed from circulation by the tissues is reduced; and (2) the rate at which glucose is added to circulation by the liver is increased.

Because transport of glucose into muscle and adipose tissue is insulin-dependent, deficiency of insulin impairs uptake of glucose by these tissues. Insulin is not required for entry of glucose into hepatic cells, but it does regulate the synthesis of the hepatic enzyme glucokinase. In the absence of insulin, the glucokinase content of the liver falls and with it the capacity of the liver to perform the first reaction of glucose metabolism. Insulin also plays a role in regulating secretion of glucagon from the pancreas. Absence of insulin leads to increased secretion of glucagon, which in turn activates both liver glycogenolysis and gluconeogenesis.

Lipid Synthesis & Transp

OBJECTIVES

- Be able to describe the synthesis of fatty acids from glucose.

- Know the principal functions of the pentose phosphate pathway.

- Be able to trace the pathway by which dietary glucose is converted to fatty acids and stored as triglyceride in adipose.

- Be able to explain how lipid metabolism and transport are regulated.

- Be able to name and describe the major classes of membrane lipids and to describe the role of nucleoside triphosphates in their synthesis.

- Be able to name the substrates and the key intermediates in the synthesis of cholesterol. Know which is the rate-limiting reaction of the pathway and how it is regulated.

- Know the names, components, and functions of the lipoprotein particles. Be able to describe where each is formed, how it delivers its contents to its target tissues, and how its remnants, if any, are removed from circulation.

LIPIDS ARE biologic compounds largely or entirely composed of nonpolar groups. As a consequence of this property, they are readily soluble in nonpolar solvents and relatively insoluble in water. Some lipids function as fuels (see Chapter 7), others as components of membranes (see Chapter 14) or as precursors of hormones (see Chapter 16), and still others in a variety of additional specialized roles. The nonpolar nature of lipids is of central

...rtance to many of these functions but poses special problems for their ...nsport throughout the body.

The pathways in which lipids participate are outlined in the chapters cited above. Here we will describe the synthesis of fuel and membrane lipids and the solution to the problem of transport.

Fatty Acid Synthesis

Most tissues are able to use lipids as fuel and do so whenever glucose is limiting. Because lipids are more reduced than other fuels and because they are stored without hydration, catabolism of lipids yields a larger amount of ATP than catabolism of an equal weight of either carbohydrate or protein. By decreasing the demand for glucose, lipids play an essential supporting role in glucose homeostasis. During feeding, a portion of the excess dietary glucose and amino acids is converted to fatty acids. Both dietary fatty acids and those synthesized from glucose are stored in adipose and other tissues as triglycerides. During fasting, fatty acids are released from storage and are used by many tissues as an alternative to glucose.

Fatty acid synthesis is performed in the cytoplasm of the liver and, to a lesser extent, of adipose tissue using acetyl-CoA derived from the partial breakdown of glucose and amino acids. The primary product of fatty acid synthesis in humans is the 16-carbon saturated fatty acid palmitate. It is synthesized 2 carbons at a time in a series of reactions catalyzed by fatty acid synthase. Although palmitate is assembled in 2-carbon increments, only carbons 15 and 16 are derived directly from acetyl-CoA. Carbons 1–14 enter the chain via malonyl-CoA, which is formed in a reaction catalyzed by acetyl-CoA carboxylase, a biotin-requiring enzyme (Fig 9–1). The ATP consumed in the production of malonyl-CoA indirectly provides the energy needed for fatty acid synthesis.

Fatty acid synthase is a complex enzyme made up of 2 identical polypeptide chains. Each polypeptide includes within its structure all of the enzymatic activities required for the synthesis of palmitate from acetyl-CoA and malonyl-CoA. The mechanism by which palmitate is synthesized is

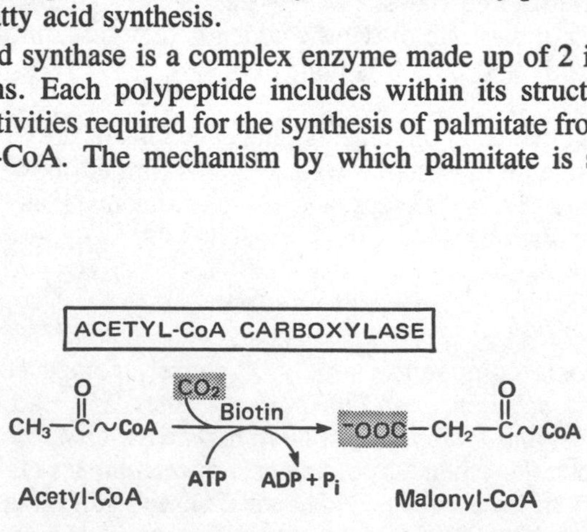

Figure 9–1. Formation of malonyl-CoA from acetyl-CoA.

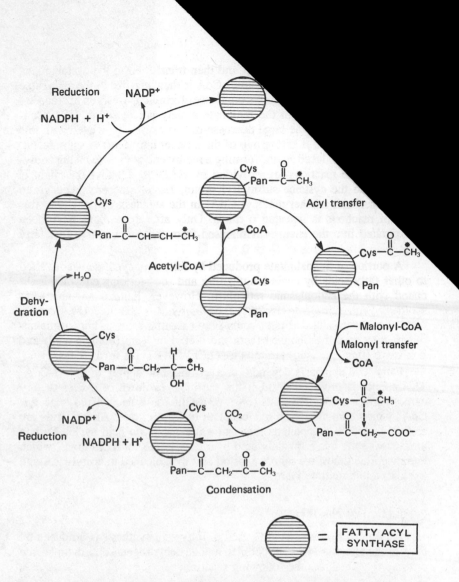

Figure 9–2. Synthesis of palmitate by fatty acid synthase. The reactive cysteine side chain is represented as –Cys and that of the phosphopantetheine prosthetic group as –Pan. (Redrawn after Löffler G et al: *Physiologische Chemie.* Springer-Verlag, 1979.)

shown schematically in Fig 9–2. During fatty acid synthesis, the substrates of the reaction become covalently attached to the enzyme through 2 sulfhydryl groups; one is part of a cysteine residue and the other part of phosphopantetheine, a prosthetic group formed from the vitamin pantothenate. In the first step of fatty acid synthesis, the acetyl group of acetyl-CoA is attached

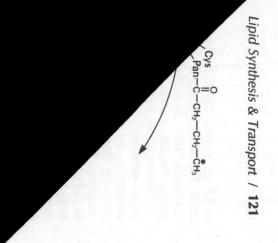

...ferred to the cysteine side ... attached to the coenzyme ...ng a 4-carbon acetoacetyl ... is released as CO_2. Con-...nergy associated with this ... fatty acid is then reduced, ...he group. Reducing equiv-...PH. Finally, the chain is ...ving the prosthetic group ...ynthesis of palmitate, this ...er 16 carbons have been ...he enzyme as a free fatty

...acid synthase is converted ...esaturating enzymes asso-...r, humans are unable to synthesize either **linoleate** (18:2;9,12) or **linolenate** (18:3;9,12,15). Because both of these unsaturated fatty acids play essential roles in human metabolism, they must be obtained from the diet. Linoleate (linoleic acid) and linolenate (linolenic acid) are thus **essential fatty acids** for humans.

Fatty acid synthesis depends on a cytoplasmic supply of acetyl-CoA. This supply is provided through the partial breakdown of excess dietary carbohydrates and amino acids (Fig 9–3). Glucose is catabolized to acetyl-CoA, which combines with oxaloacetate to form citrate. Amino acids are also degraded to compounds that can enter the citric acid cycle. Carbon atoms leave the cycle for fatty acid synthesis in the form of citrate, which is transported out of the mitochondrion and then cleaved to acetyl-CoA and oxaloacetate by citrate lyase.

Production of NADPH

A large part of the NADPH used in fatty acid synthesis is produced by the **"malic enzyme"** (NADP-coupled malate dehydrogenase), a cytoplasmic enzyme that catalyzes the following reaction:

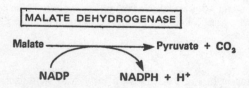

A second major source of NADPH is the **pentose phosphate pathway,** also known as the **hexose monophosphate shunt.** This pathway produces 2 substrates for anabolism: NADPH and ribose 5-phosphate. The pathway

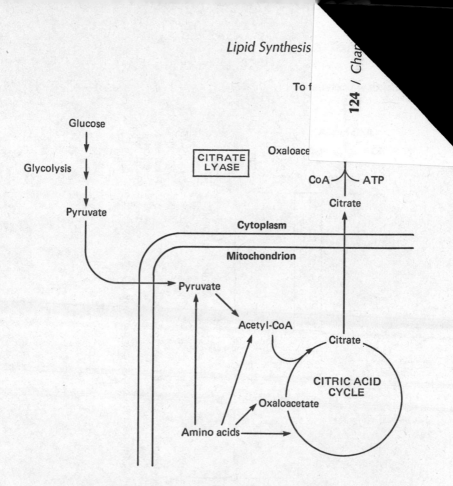

Figure 9–3. The origin of the substrates for fatty acid synthesis.

has 2 independently regulated parts, one oxidative and the other nonoxidative. In the oxidative phase, glucose 6-phosphate is reduced and decarboxylated to yield NADPH, ribulose 5-phosphate, and CO_2 (Fig 9–4). In the nonoxidative phase, ribulose 5-phosphate is further metabolized to form ribose 5-phosphate and compounds that can enter glycolysis. Because it includes a decarboxylation reaction, the oxidative phase of the pentose phosphate pathway is irreversible, whereas the second phase operates freely in both directions. The rate-limiting step in the pathway is that catalyzed by glucose-6-phosphate dehydrogenase (G6PD). Synthesis of G6PD is induced during feeding and repressed during fasting.

Several human genetic variants of G6PD, some of which significantly impair NADPH production, have been described. Individuals with G6PD deficiency develop hemolytic anemia when exposed to certain agents, including a component of fava beans and some drugs (aspirin, sulfonamides, and the antimalarial primaquine). In these cases, hemolysis is caused by hydrogen peroxide oxidizing and thereby damaging cell membranes and

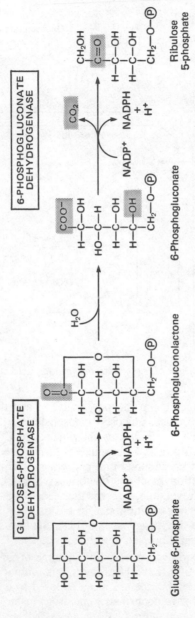

Figure 9–4. The oxidative portion of the pentose phosphate pathway. $\circled{P}$ = phosphoryl group.

other cell components. In red blood cells of normal individuals, hydrogen peroxide is reduced at the expense of glutathione, a tripeptide, and NADPH (Fig 9–5). The pentose phosphate pathway is the only source of NADPH in red blood cells, and the damage caused by hydrogen peroxide is therefore increased in individuals with G6PD deficiency.

The nonoxidative portion of the pentose phosphate pathway includes an enzyme known as transketolase. Transketolase catalyzes 2 reversible reactions in which carbons 1 and 2 of a ketose are transferred to the aldehyde carbon of an aldose (Fig 9–6). Thiamin pyrophosphate serves as a coenzyme in these reactions. A genetic variant of transketolase has been implicated in the Wernicke-Korsakoff syndrome, frequently encountered in alcoholics. Symptoms of this syndrome include paralysis of eye movements, abnormal gait and stance, and deranged mental function. The transketolase of some Wernicke-Korsakoff patients has a reduced affinity for its coenzyme. Onset of symptoms is precipitated by dietary deficiency of thiamin (vitamin B_1), common among alcoholics. The condition can be ameliorated by supplementing the diet with vitamin B_1. It is not yet apparent why a limitation in transketolase activity results in neuropsychiatric symptoms.

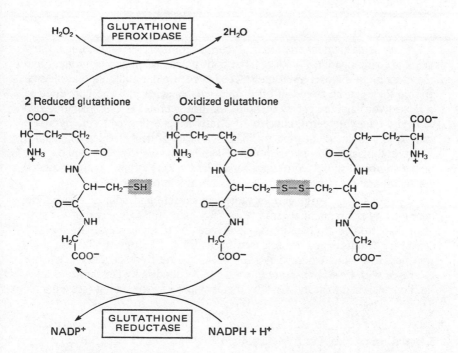

Figure 9–5. In the red blood cell, hydrogen peroxide is reduced at the expense of glutathione and NADPH.

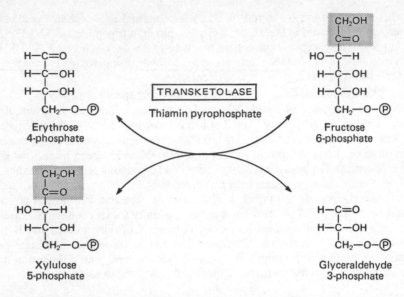

Figure 9–6. One of 2 reactions catalyzed by transketolase. Ⓟ = phosphoryl group.

Fatty acids are converted to triglycerides for transport and storage. Fatty acids are amphipathic (ie, they have both polar and nonpolar groups) and therefore align themselves at interfaces between polar and nonpolar phases. Before they can be stored in oil droplets, fatty acids must be converted to a more hydrophobic species of lipid, the triglyceride. A **triglyceride,** also termed a **triacylglycerol,** consists of 3 fatty acids esterified to the hydroxyl groups of a molecule of glycerol. The synthesis of triglycerides takes place in the cytoplasm. The major route of synthesis uses glycerol 3-phosphate as the source of the glycerol backbone (Fig 9–7). Fatty acids are activated for incorporation into triglycerides by attachment to CoA.

During fasting, triglycerides stored in muscle and adipose tissues are degraded by the sequential removal of fatty acid units. Removal of the first fatty acid from a triglyceride is catalyzed by triacylglycerol lipase. The remaining 2 fatty acids are released by additional lipases. Following their transport to tissues where they will be used, the released fatty acids are esterified with CoA, transported into the mitochondria by the carnitine shuttle, and degraded to acetyl-CoA by the enzymes of β-oxidation (see Chapter 7).

Regulation of Fuel Lipid Metabolism

Fuel lipids are synthesized and stored when glucose is abundant and are catabolized when glucose is scarce, under conditions of stress, and during

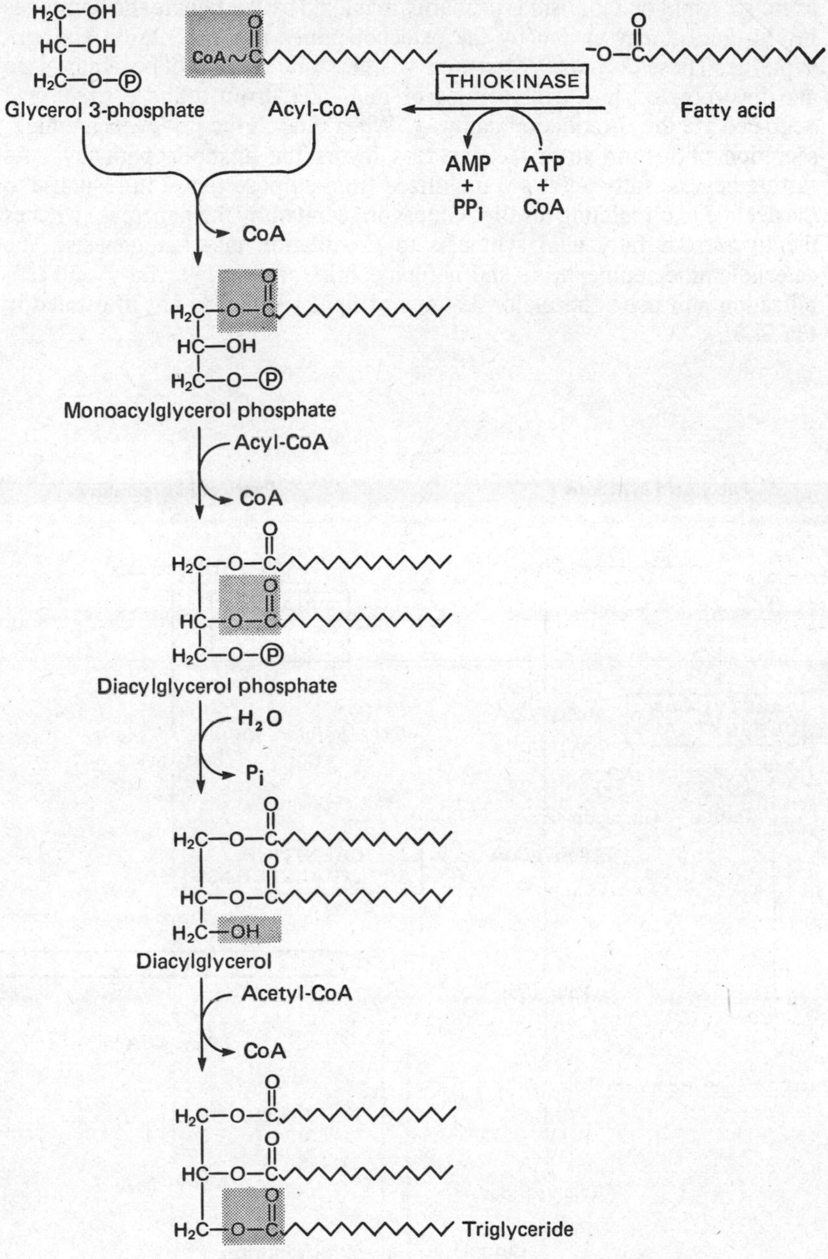

Figure 9–7. Fatty acids are incorporated into triglycerides for storage and transport. $\textcircled{P}$ = phosphoryl group.

exercise. Lipid metabolism is primarily regulated by the pancreatic hormones insulin and glucagon and by the catecholamines norepinephrine and epinephrine. These hormones determine whether fatty acids will be synthesized and incorporated into triglycerides or mobilized from adipose tissue and degraded via the β-oxidation pathway. When blood glucose levels are high, secretion of insulin from the pancreas favors the anabolic pathways. As fasting begins, fatty acids are mobilized from adipose tissue in response to the decline in circulating insulin. Glucagon secreted by the pancreas switches the liver from fatty acid synthesis to β-oxidation and ketogenesis. The catecholamines epinephrine and norepinephrine also trigger fatty acid mobilization and use. The major features of lipid regulation are illustrated in Fig 9–8.

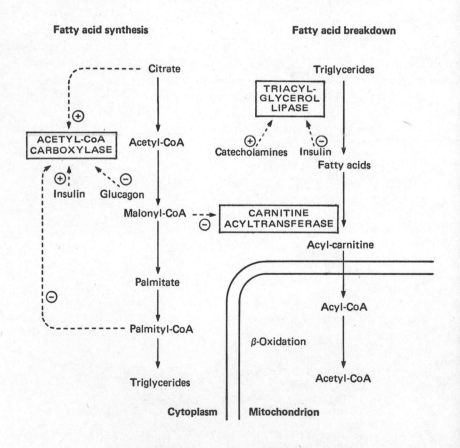

Figure 9–8. Regulation of fuel lipid metabolism.

The switch between synthesis and breakdown of triglycerides in adipose tissue is controlled at the level of triacylglycerol lipase, which catalyzes the first step in the mobilization of fatty acids. **Triacylglycerol lipase,** also known as hormone-sensitive lipase, is activated by phosphorylation and inactivated by dephosphorylation. Phosphorylation and dephosphorylation are controlled by hormonal signals. During fasting, triacylglycerol lipase is phosphorylated by cAMP-dependent protein kinase and thereby activated. The major stimulus for cAMP production in adipose may be norepinephrine released by the sympathetic nervous system. Epinephrine released from the adrenals in time of stress also promotes synthesis of cAMP and mobilization of fatty acids. Dephosphorylation is favored when blood glucose and insulin levels are elevated. The balance between synthesis and breakdown of fatty acids is controlled via the rate-limiting enzyme of fatty acid synthesis, **acetyl-CoA carboxylase.** During fasting, when glucagon levels are elevated, acetyl-CoA carboxylase is phosphorylated, and thereby inhibited, by cAMP-dependent protein kinase. When glucose is abundant, the presence of insulin leads to dephosphorylation and activation of the enzyme. Acetyl-CoA carboxylase is also allosterically activated by citrate and inhibited by palmityl-CoA. The latter control limits the production of fatty acids to the amount that can be stored or used in the synthesis of other lipids.

The rate at which fatty acids are degraded in β-oxidation is limited by the rate at which they are transported into the mitochondria by the carnitine shuttle. **Carnitine acyltransferase I,** the enzyme that catalyzes the formation of acylcarnitine on the cytoplasmic side of the mitochondrial membrane, is inhibited by the product of acetyl-CoA carboxylase, malonyl-CoA. Thus, activation of the pathway for fatty acid synthesis simultaneously shuts off the pathway for fatty acid catabolism.

Because most of the substrates for ketogenesis are provided by the catabolism of fatty acids in the liver, regulation of ketogenesis is closely tied to regulation of fatty acid metabolism. The rate of ketogenesis parallels the rate at which fatty acids are mobilized from adipose tissue and thus is increased during fasting.

Because insulin plays a major role in regulating the metabolism of lipids and keto acids, this regulation is severely deranged in untreated type I diabetes mellitus. Absence of insulin leads to rapid mobilization of fatty acids from adipose tissue and rapid synthesis of· keto acids. The resulting increase in the plasma concentration of keto acids causes the pH of the plasma to fall **(ketoacidosis).**

Membrane Lipids

Cell membranes provide a hydrophobic barrier around the aqueous compartments of the cell and prevent diffusion of water-soluble compounds from one compartment to another. Membranes are largely composed of amphipathic lipids that constitute the major part of both the hydrophobic barrier

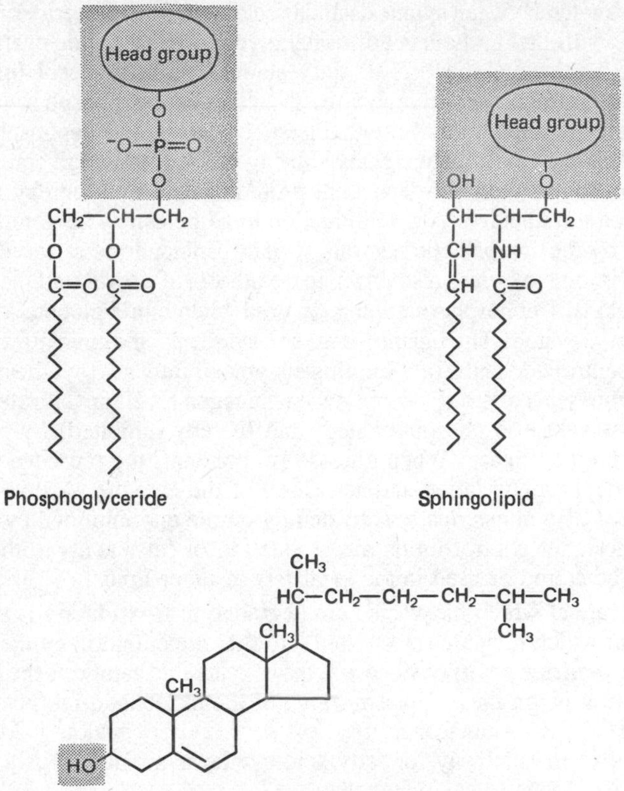

Phosphoglyceride   Sphingolipid

Cholesterol

Figure 9-9. The structures of phosphoglycerides, sphingolipids, and cholesterol. The polar portion of each molecule is shaded.

and the polar surface of the membrane. Membranes contain 3 classes of amphipathic lipids: phosphoglycerides, sphingolipids, and cholesterol (Fig 9–9).

Phosphoglycerides contain 2 fatty acids joined in ester linkages to a glycerol backbone. Also attached to the glycerol and forming the polar portion of the lipid is a phosphoryl group and one of several head groups—ethanolamine ($-CH_2-CH_2-NH_2$), choline ($-CH_3-CH_3-N-[CH_3]_3$), inositol (a 6-carbon sugar), or serine. Sphingolipids also contain 2 fatty acids and a polar head group. However, the sphingolipid backbone is derived not from glycerol but from serine. The α-amino group of the amino acid forms the linkage to one of the substituent fatty acids. The second fatty acid forms an alcohol upon attachment to the serine backbone. The polar head groups of

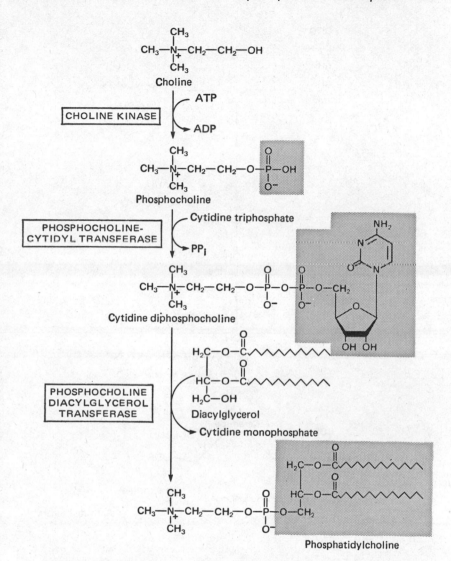

Figure 9–10. Synthesis of phosphatidylcholine. CTP is used to activate the phosphorylcholine head group for attachment to diacylglycerol.

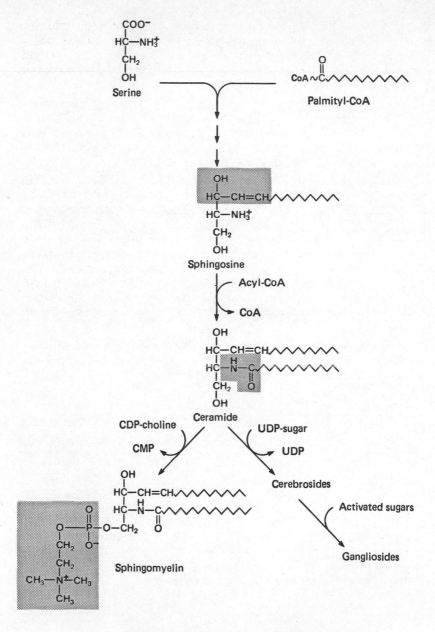

Figure 9–11. Sphingolipid synthesis.

the sphingolipids are formed from either phosphorylcholine (in sphingo-myelin) or carbohydrates (in glycolipids). Cholesterol differs from the phos-phoglycerides and the sphingolipids in that it is a planar ring compound and does not contain fatty acids. A hydroxyl group constitutes the polar portion of the cholesterol molecule.

Phosphoglycerides are synthesized from either phosphatidic acid or diac-ylglycerol. A nucleotide, cytidine triphosphate (CTP), participates in their synthesis by activating phosphatidic acid or the polar head group that is to be added. A representative pathway, that by which phosphatidylcholine is synthesized, is shown in Fig 9–10.

Sphingolipid synthesis begins with the formation of **sphingosine** (Fig 9–11) from serine and palmitate. Sphingosine is converted to **ceramide** by the addition of a second fatty acid. The sphingolipids are completed by the addition of the appropriate head group. Both cytidine and uridine nucleotides are used to activate the head groups for addition. Four types of carbohydrate units are found in the glycolipid head groups: glucose, galactose, neuraminic acid (a 9-carbon negatively charged sugar), and N-acetylgalactosamine. The head structure may contain as few as one or as many as 9 carbohydrate units. Because they are abundant in nervous tissue, those glycolipids that contain neuraminic acid and N-acetylgalactosamine have been termed gan-gliosides.

A number of diseases, called **sphingolipidoses,** are caused by defi-ciencies of enzymes that mediate the breakdown of glycolipids (Table 9–1).

Table 9–1. Sphingolipidoses.

Disease	Enzyme Deficiency	Lipid Accumulating*†	Clinical Symptoms
Fabry's disease	Ceramide trihexosidase	Cer—Glc—Gal⌡Gal	Kidney failure Skin symptoms
Gaucher's disease	Glucocerebrosidase	Cer⌡Glc	Mental retardation Enlarged liver and spleen
Krabbe's disease	Galactocerebrosidase	Cer⌡Gal	Mental retardation
Metachromatic leukodystrophy	Sulfatidase	Cer—Gal⌡OSO₃	Mental retardation
Niemann-Pick disease	Sphingomyelinase	Cer⌡℗—choline (sphingomyelin)	Mental retardation Enlarged liver and spleen
Tay-Sachs disease	Hexosaminidase A	Cer—Glc—Gal⌡GalNAc | NANA	Mental retardation Blindness Muscular weakness

*⌡ = Site of deficient enzymatic reaction.
†Cer, ceramide; Glc, glucose; Gal, galactose; GalNAc, N-acetyl galactosamine; NANA, neuraminic acid.

Individuals suffering from lipidosis accumulate abnormally high amounts of the lipid substrate for the enzyme that is defective. Mental retardation is common in many of these diseases.

In humans, cholesterol is both acquired from the diet and synthesized de novo, primarily in the liver. An overview of cholesterol biosynthesis is shown in Fig 9–12. In the first part of the pathway, 3 acetyl-CoA units are

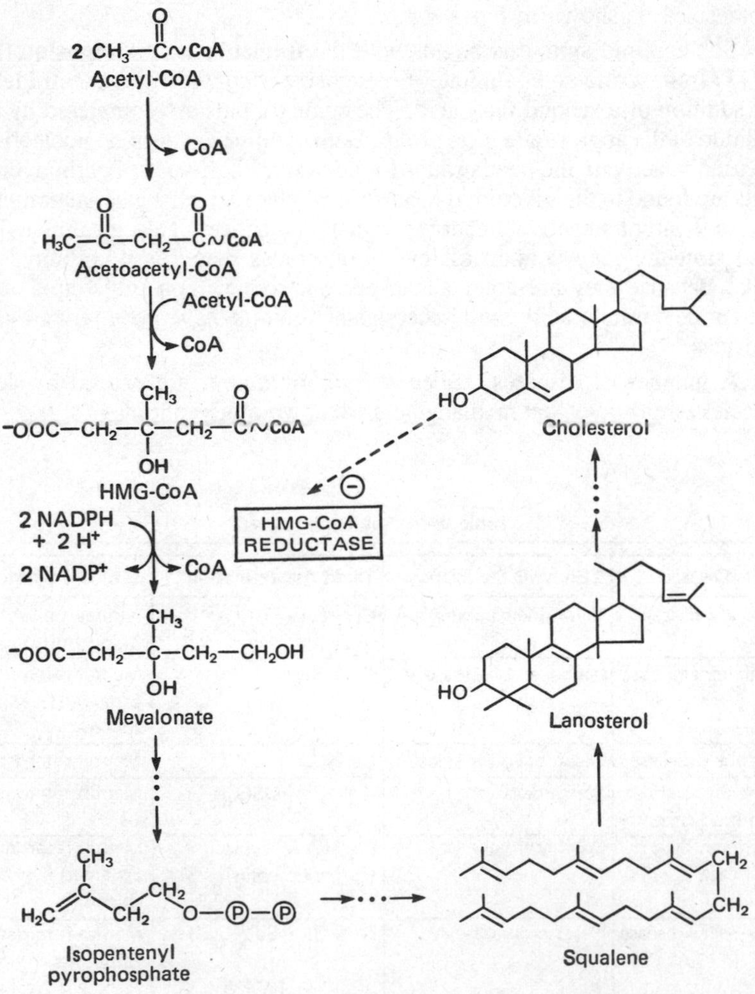

Figure 9–12. Synthesis of cholesterol. The rate-limiting enzyme, HMG-CoA reductase, is feedback-inhibited by cholesterol. P = phosphoryl group. The notation $\rightarrow \cdots \rightarrow$ indicates that several intervening steps are not shown.

used to synthesize β-hydroxy-β-methylglutaryl-CoA (HMG-CoA). Recall that HMG-CoA is also an intermediate in ketogenesis (see Chapter 7). Because cholesterol biosynthesis takes place in the cytoplasm and ketogenesis in the mitochondrion, the 2 pathways do not interact through this common intermediate. HMG-CoA formed in the cholesterol pathway is reduced and detached from CoA, forming mevalonate. The latter is then converted to an activated isoprene compound, isopentenyl pyrophosphate. Six isoprene units condense to form squalene, which is then cyclized to lanosterol. The latter is modified in several steps to form cholesterol. HMG-CoA reductase, the rate-limiting enzyme in cholesterol biosynthesis, is feedback-inhibited by cholesterol. Thus, when dietary cholesterol is available in amounts adequate for membrane and sterol synthesis, the pathway is inhibited.

Like fatty acids, cholesterol is converted to a more hydrophobic form of lipid before being transported or stored. Cholesterol is made more hydrophobic by esterification with a fatty acid, forming a cholesteryl ester. Within cells, esterification is catalyzed by acyl-cholesterol acyltransferase (ACAT), which uses fatty acyl-CoA as the source of the fatty acid (Fig 9–13).

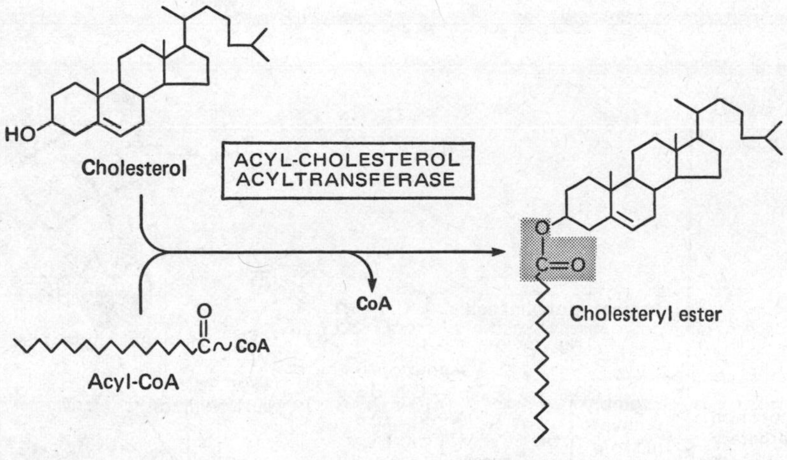

Figure 9–13. Esterification of cholesterol by acyl-cholesterol acyltransferase. The ester linkage is shaded.

Interorgan Transport of Lipids

Because lipids are nonpolar, they must be transported in the circulatory system by water-soluble vehicles. Lipids are carried both by a protein, serum albumin, and by aggregates of lipids and proteins known as lipoprotein particles.

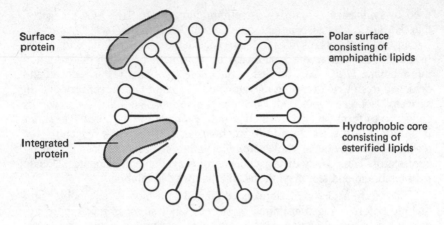

Figure 9–14. General structure of a lipoprotein particle.

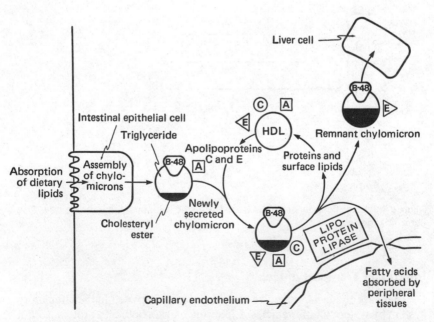

Figure 9–15. Chylomicrons deliver dietary triglycerides to peripheral tissues and cholesterol to the liver.

Free fatty acids are transported by serum albumin, a protein specialized for the transport of several nonpolar substances. Albumin carries a portion of the fatty acids absorbed from the intestine to adipose and muscle tissue and those released from adipose stores to muscle and liver. Serum albumin also transports bilirubin (see Chapter 11) and lysolecithin (see Fig 9–18).

Triglycerides and cholesteryl esters are transported by several types of lipoprotein particles, all of which have a structure like that shown in Fig 9–14. The lipids to be transported are contained within the nonpolar core. Water solubility is conferred on the particle by a surface monolayer of amphipathic lipids (cholesterol and phospholipids) oriented with their polar groups to the outside. A number of proteins, called apolipoproteins, are attached to the surface or integrated into the particle. These proteins mediate (1) the secretion of the particles from the cells in which they are formed, (2) the delivery of the core material to the target tissue, and (3) the catabolism of the remnants that remain when the particles are exhausted of their contents. There are 4 major classes of lipoprotein particles. Their functions and structures are summarized in Table 9–2 and described below.

Table 9–2. Structures and functions of the major classes of lipoprotein particles.

Name	Function and Route	Core Lipids	Apolipoproteins and Their Functions
Chylomicrons	Deliver dietary triglycerides to peripheral tissues and dietary cholesterol to liver. Secreted by intestinal epithelial cells. Deliver triglycerides via lipoprotein lipase. Remnant removed by liver.	Primarily triglycerides, some cholesteryl esters	B-48 mediates secretion. A is required for formation of new HDL. C activates lipoprotein lipase. E mediates uptake of remnant by liver.
VLDL	Deliver hepatic triglycerides to peripheral tissues. Secreted by liver. Deliver triglycerides via lipoprotein lipase. Majority of remnants converted to LDL.	Primarily triglycerides, some cholesteryl esters	B-100 mediates secretion. C activates lipoprotein lipase. E mediates metabolism of remnant.
LDL	Deliver hepatic cholesterol to peripheral tissues. Formed at the liver from remnants of VLDL. Taken up by target cells via receptor-mediated endocytosis.	Cholesteryl esters	B-100 mediates binding to cell surface receptor for endocytosis.
HDL	Mediate centripetal transport of cholesterol. Some derived from the surface components of chylomicrons. Others secreted by the liver.	Cholesteryl esters	LCAT catalyzes esterification of cholesterol. A activates LCAT. D mediates transfer of cholesteryl esters to other lipoprotein particles.

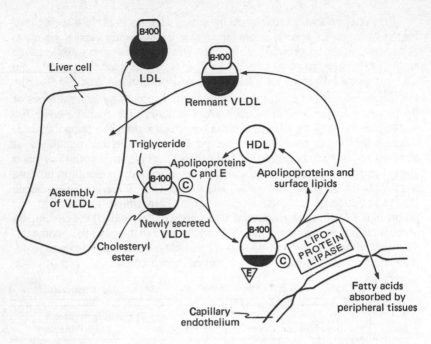

Figure 9–16. VLDL transport triglycerides from the liver to adipose and muscle.

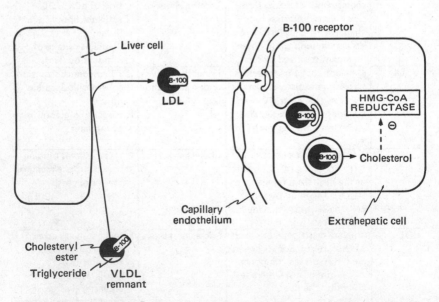

Figure 9–17. LDL deliver cholesterol to peripheral tissues via receptor-mediated endocytosis.

Chylomicrons transport dietary lipids absorbed from the intestine to the liver and peripheral tissues (Fig 9–15). Lipids taken in as food are digested in the intestine to free fatty acids, cholesterol, and β-monoglycerides (single fatty acids esterified to the β-hydroxyl group of glycerol). These products are absorbed by the intestinal epithelial cells. Short-chain fatty acids are released to the circulation to be transported by albumin. The other lipids are converted within the intestinal cells to triglycerides and cholesteryl esters and are packaged into chylomicrons for transport. Newly formed chylomicrons are released into the lymph, carrying apolipoprotein B-48, which mediates secretion of the particle, and the A apolipoproteins. Chylomicrons are carried by the lymph into the bloodstream. In circulation, they interact with **high-density lipoprotein particles (HDL)**, gaining C and E apolipoproteins and giving up surface lipids.

As chylomicrons circulate past peripheral tissues, including adipose tissue and muscle, triglycerides contained in the core are degraded, and the fatty acids released in the process are absorbed by neighboring cells. Breakdown of chylomicron triglycerides is catalyzed by **lipoprotein lipase,** which is bound to the surface of the capillary endothelium. Humans have 2 forms of lipoprotein lipase. That found in the capillaries of muscle has a low K_m for triglycerides and is not subject to hormonal regulation, whereas that of adipose has a high K_m and is induced by insulin. The differences in the catalytic properties and regulation of the 2 forms of lipoprotein lipase ensure that muscle has preferential access to fatty acids when they are scarce. Both forms of lipoprotein lipase are activated by apolipoprotein C-II, carried on the chylomicron. Within 15 minutes after its secretion, the chylomicron is largely depleted of its triglycerides, and with the help of apolipoprotein E, the remnant particle is taken up and catabolized by the liver. During the circulation of the chylomicron, cholesteryl esters are transferred to its core from HDL. When the remnant is catabolized, these as well as the cholesteryl esters derived from the diet are added to the pool of hepatic cholesterol. If adequate cholesterol is delivered to the liver by chylomicron remnants, hepatic HMG-CoA reductase is inhibited, thereby limiting the endogenous synthesis of cholesterol.

Using fatty acids synthesized de novo by the liver plus those taken up from circulation, the liver synthesizes triglycerides and incorporates them into another class of lipoprotein particle, the **very low density lipoproteins (VLDL)** (Fig 9–16). When secreted from the liver, VLDL carry apolipoproteins B-100 and C-I, C-II, and C-III. Although smaller, VLDL are similar to chylomicrons in many respects. Their secretion depends on the presence of a B apolipoprotein (B-100); in circulation they exchange apolipoproteins and surface components with HDL; and they are substrates for lipoprotein lipase. Thus, like the chylomicrons, VLDL deliver fatty acids to adipose and muscle. VLDL remnants are metabolized by the liver; some are degraded, but the majority are converted to **low-density lipoprotein particles (LDL).**

LDL, which are the major cholesterol-carrying lipoprotein particles in the circulation, deliver cholesterol from the liver to peripheral cells (Fig

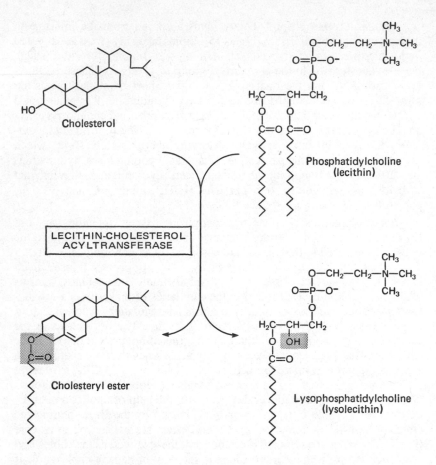

Figure 9–18. Lecithin-cholesterol acyltransferase catalyzes the formation of cholesteryl ester using phosphatidylcholine (lecithin) as the acyl group donor.

9–17). In the process by which remnants of VLDL are converted to LDL, the particles are stripped of any remaining triglycerides and all of the apolipoproteins except B-100. Additional cholesteryl esters are added to LDL from the hepatic pool. After emerging from the liver, newly formed LDL circulate to the peripheral tissues. Unlike the chylomicrons and VLDL, LDL are small enough to pass through pores in the capillary endothelium and leave the circulatory system. LDL deliver their contents to peripheral cells by a process termed **receptor-mediated endocytosis.** This process is mediated by apolipoprotein B-100 and a receptor protein located on the surface of the peripheral cell. Endocytosis is initiated by binding of apolipoprotein B-100 to the receptor protein. The cell membrane subsequently invaginates, forming a cytoplasmic vesicle. The vesicle then fuses with a lysosome, and

LDL contained inside are degraded by lysosomal enzym(
carried in the core of LDL are released as free chole؟
used immediately for the construction of new membrar
be esterified by acyl:cholesterol acyltransferase for st(
livered to cells by LDL contributes to the total cellula
inhibits cholesterol biosynthesis in those cells.

HDL form a heterogeneous group of particles that participate in a process termed the **centripetal transport** of cholesterol. In this process, HDL transport excess cholesterol away from peripheral tissues. There are at least 2 routes by which HDL are synthesized. Some are formed in circulation from surface components carried on chylomicrons. Others are secreted by the liver.

In centripetal transport, excess cholesterol is transferred from peripheral cell membranes into the surface monolayers of HDL. HDL convert free cholesterol to cholesteryl esters in a reaction catalyzed by **lecithin-cholesterol acyltransferase (LCAT)** (Fig 9–18). A protein carried by HDL, apolipoprotein A-II, is required for activation of LCAT. The fatty acyl group used to esterify cholesterol of HDL is donated by phosphatidylcholine (lecithin) taken from the surface layer of HDL. Cholesteryl esters formed in the LCAT reaction are transferred from HDL to other lipoprotein particles, including LDL, VLDL, and chylomicrons. Remnants of chylomicrons and VLDL are taken up by the liver, and thus the cholesterol removed from peripheral tissues is added to the hepatic cholesterol pool.

Amino Acid Metabolism

OBJECTIVES

- Be able to describe the role of pyridoxal phosphate in amino acid metabolism.

- Be able to name the essential amino acids.

- Know the physiologic importance of urea synthesis. Be able to describe the steps in the pathway by which the amino groups of amino acids are incorporated into urea. Be able to explain the regulation of this process.

- Know what is meant by the terms "ketogenic" and "glucogenic." Know which amino acids are ketogenic and which are glucogenic and why.

- Be able to describe the genetic basis of phenylketonuria. Know how phenylketonuria affects dietary requirements for amino acids.

AMINO ACIDS have 3 major roles in metabolism. They serve as substrates for the synthesis of protein, provide nitrogen for the synthesis of other nitrogen-containing compounds, and are catabolized as fuels. Because amino acids serve as the major source of nitrogen for anabolic pathways, they must be continuously available for metabolism. However, unlike carbohydrates and lipids, amino acids cannot be stored by the body for future use and therefore must be supplied by the diet on a regular basis.

142

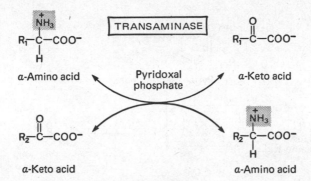

Figure 10–1. A transaminase catalyzes the transfer of the α-amino group of an amino acid to the α carbon of a keto acid.

Pyridoxal Phosphate

Both the anabolic and the catabolic pathways in amino acid metabolism include **transamination reactions,** ie, reactions involving the transfer of the α-amino group of an amino acid to the α carbon of a keto acid, thereby forming a new amino acid and a new keto acid (Fig 10–1). Transamination reactions are catalyzed by a variety of transaminases (also called aminotransferases), all of which employ a common mechanism and utilize a coenzyme derived from vitamin B_6, pyridoxal phosphate. In a transamination reaction, pyridoxal phosphate acts as an intermediate carrier of the amino group that is being transferred (Fig 10–2). Pyridoxal phosphate also serves as a coenzyme in a number of reactions other than transaminations. (See p 153 and Fig 11–2.)

Essential & Nonessential Amino Acids

In every tissue, proteins are continually being synthesized and degraded. Protein synthesis can proceed only when there are adequate pools of the 20 common amino acids (see Table 3–1). Humans do not have the ability to synthesize 10 of the necessary 20 amino acids and must obtain them from the diet. These 10 are termed the nutritionally essential amino acids (Table 10–1). The 10 nonessential amino acids—those that can be synthesized by humans—are formed by 3 general mechanisms: assimilation of free ammonia, transamination, and modification of the carbon skeletons of existing amino acids.

Two of the nonessential amino acids, **glutamate** and **glutamine,** can be synthesized via the assimilation of free ammonia (Fig 10–3). Formation of glutamate from ammonia and α-ketoglutarate is catalyzed by glutamate

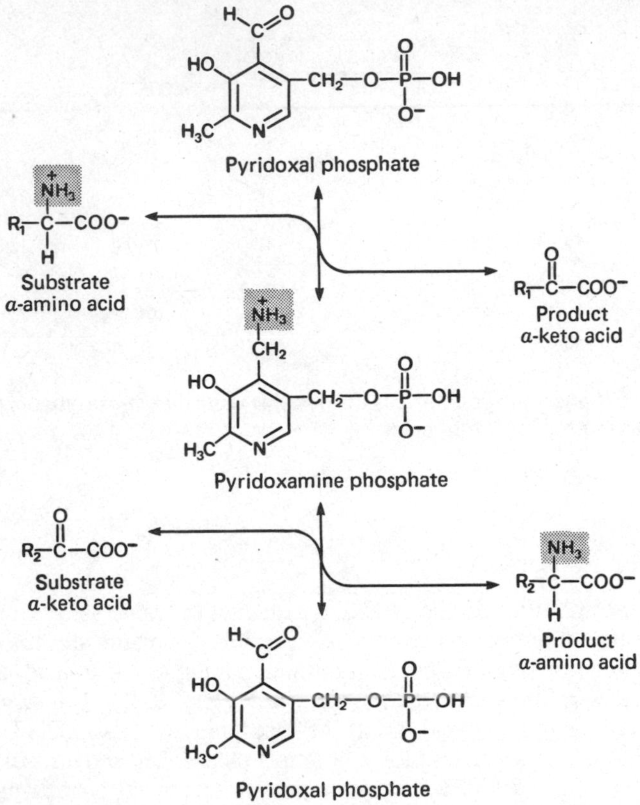

Figure 10–2. Pyridoxal phosphate acts as an intermediate carrier of amino groups in transamination reactions.

Table 10–1. Nutritionally essential and nonessential amino acids.

Nutritionally Essential	Nutritionally Nonessential
Arg	Ala
His	Asp
Ile	Asn
Leu	Cys
Lys	Glu
Met	Gln
Phe	Gly
Thr	Pro
Trp	Ser
Val	Tyr

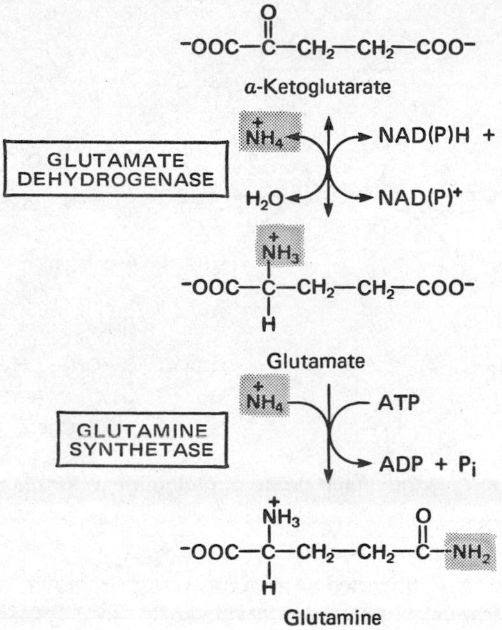

$$-OOC-\overset{\overset{\displaystyle O}{\|}}{C}-CH_2-CH_2-COO^-$$

α-Ketoglutarate

GLUTAMATE
DEHYDROGENASE

$\overset{+}{NH_4}$ NAD(P)H + H$^+$

H$_2$O   NAD(P)$^+$

$$-OOC-\overset{\overset{\displaystyle +}{\overset{\displaystyle NH_3}{|}}}{\underset{\underset{\displaystyle H}{|}}{C}}-CH_2-CH_2-COO^-$$

Glutamate

GLUTAMINE
SYNTHETASE

$\overset{+}{NH_4}$ ATP

ADP + P$_i$

$$-OOC-\overset{\overset{\displaystyle +}{\overset{\displaystyle NH_3}{|}}}{\underset{\underset{\displaystyle H}{|}}{C}}-CH_2-CH_2-\overset{\overset{\displaystyle O}{\|}}{C}-NH_2$$

Glutamine

Figure 10–3. Glutamate and glutamine can be formed by the assimilation of free ammonia. The abbreviation NAD(P)H indicates that both NADH and NADPH can serve as the source of reducing equivalents.

dehydrogenase. This reaction is reversible and plays a role in both synthesis and breakdown of glutamate. Both NADPH and NADH can serve as the source of reducing equivalents used in this reaction. Glutamine synthetase catalyzes the ATP-dependent formation of glutamine, using glutamate and ammonia as substrates. Most of the ammonia used in these reactions is formed through the deamination of amino acids (see below).

Almost all amino acids can participate in transamination reactions, and because these reactions are thermodynamically reversible, they theoretically could be used for the synthesis of most of the 20 amino acids. In reality, however, the number synthesized by this route is limited by the availability of the α-keto acids corresponding to the amino acid carbon skeleton. **Alanine, glutamate,** and **aspartate** can be synthesized by transamination because the corresponding α-keto acids (pyruvate, α-ketoglutarate, and oxaloacetate) are produced through the metabolism of fuels. **Serine** is synthesized by the transamination and dephosphorylation of 3-phosphoglycerate, an intermediate of glycolysis.

Five amino acids are formed by the modification of the carbon skeletons of other amino acids. **Cysteine** contains atoms donated by both methionine

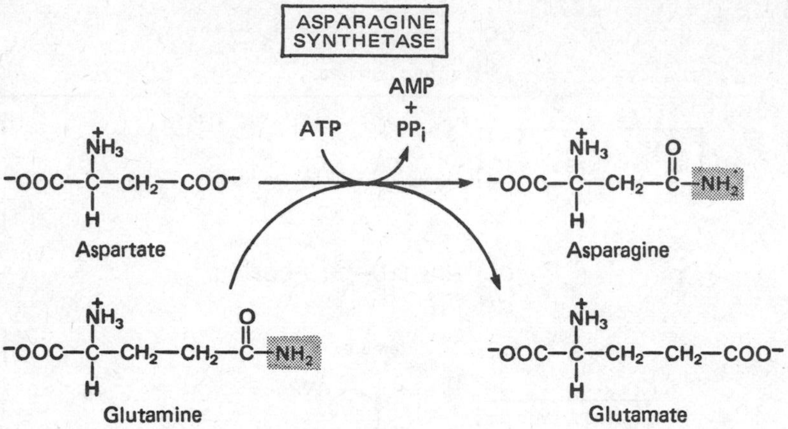

Figure 10–4. Aspartate accepts the amide group of glutamine to form asparagine.

and serine. Serine is also converted to **glycine** by the removal of its hydroxymethyl group. Phenylalanine is hydroxylated to form **tyrosine**. Glutamate is reduced and cyclized to form **proline**. The pathways for cysteine and tyrosine biosynthesis are discussed later in this chapter in connection with defects of amino acid metabolism. **Asparagine** is synthesized by the transfer of the amide group of glutamine to the β-carboxyl group of aspartate (Fig 10–4).

Amino Acid Catabolism

When amino acids in the diet exceed the requirements for protein synthesis and other anabolic pathways, the excess is catabolized for ATP production or converted to substrates for fatty acid synthesis. Cellular proteins are also turned over (broken down and replaced) at a characteristic rate of approximately 400 g/d for an adult. One-quarter of the amino acids released in this process are catabolized, along with the excess dietary amino acids. The remainder is repolymerized into proteins. Although protein is not primarily a fuel storage polymer, during starvation it is treated as such and is broken down to provide energy and substrates for gluconeogenesis and ketogenesis.

Clearly, only the carbon skeletons and not their amino groups can be used in the above-named processes. Thus, a major feature of amino acid catabolism is nitrogen disposal. Because it is extremely toxic to the brain, ammonia cannot serve as the major excretion vehicle of nitrogen. The plasma concentration of ammonia is normally maintained at a low level, approximately 100 μg/dL. Higher circulating ammonia levels result in ammonia intoxication leading eventually to coma and death. The metabolic basis for ammonia toxicity has not yet been determined. Most of the amino groups are ultimately excreted from the body in the form of **urea,** a nontoxic, water-soluble compound whose sole function in metabolism is nitrogen excretion.

$$H_2N-\overset{\overset{\displaystyle O}{\|}}{C}-NH_2$$
Urea

The amino groups of amino acids enter the pathway for urea synthesis in the form of 2 nitrogen-containing compounds, aspartate and ammonia (Fig 10–5). Through a series of transamination reactions, the amino groups of most amino acids are transferred to α-ketoglutarate, forming glutamate. A portion of the glutamate so formed serves as the donor in another transamination reaction that produces aspartate. Additional glutamate is oxidatively deaminated by glutamate dehydrogenase (mentioned earlier in the synthesis of glutamate) to provide the major source of ammonia for urea synthesis. Serine, histidine, threonine, and asparagine can also be deaminated directly.

The liver and, to a lesser extent, muscle, skin, blood cells, brain, and kidney contain the enzymes necessary for the synthesis of urea. The reactions catalyzed by these enzymes form a cycle that consumes CO_2, ammonia,

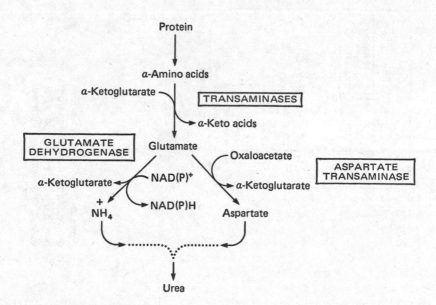

Figure 10–5. Amino groups destined for urea synthesis are collected in the form of aspartate and ammonia.

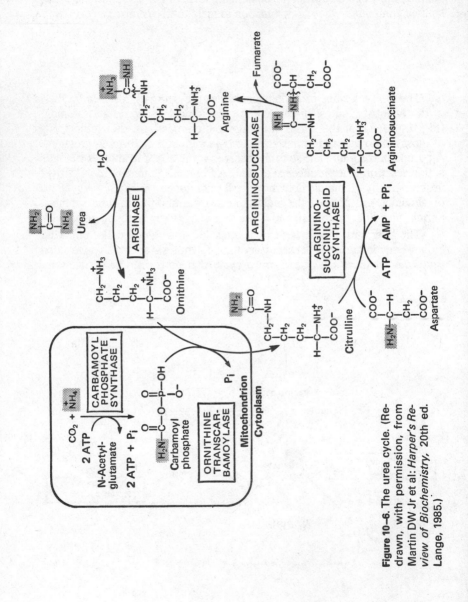

Figure 10–6. The urea cycle. (Redrawn, with permission, from Martin DW Jr et al: *Harper's Review of Biochemistry*, 20th ed. Lange, 1985.)

aspartate, and ATP (Fig 10– 6). Ammonia enters the cycle by combining with CO_2 and ATP to form carbamoyl phosphate. Because this reaction consumes 2 ATP, it is essentially irreversible and ensures the effective removal of ammonia from metabolism. Carbamoyl phosphate combines with ornithine to produce citrulline. The synthesis of carbamoyl phosphate and its combination with ornithine take place within the mitochondrial matrix. The remaining reactions in urea synthesis are performed in the cytoplasm. Aspartate, carrying the second nitrogen atom of urea, enters the cycle by condensing with citrulline to form argininosuccinate. Another ATP is consumed at this step. Argininosuccinate is cleaved to fumarate and arginine. Arginine is further hydrolyzed to yield urea and regenerate the ornithine needed for the next round of the cycle.

The activity of the urea cycle is regulated by the availability of ammonia, by allosteric control, and by enzyme induction. Glutamate dehydrogenase,

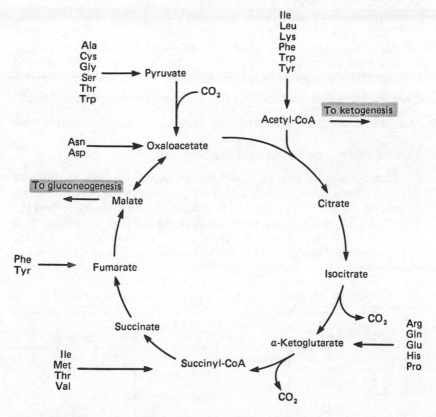

Figure 10–7. The carbon skeletons of amino acids are catabolized to pyruvate, acetyl-CoA, and intermediates of the citric acid cycle.

which provides the bulk of the ammonia for urea synthesis, is allosterically activated by ADP and GDP and inhibited by ATP and GTP and thus is regulated in response to the energy needs of the cell.

The activity of carbamoyl phosphate synthase I is allosterically regulated by a positive effector, N-acetylglutamate. This activator is synthesized from glutamate and acetyl-CoA when amino acids are abundant in the liver, a condition that coincides with heavy use of the urea cycle. On a longer time scale, the levels of urea cycle enzymes are adjusted to the nitrogen load placed on the liver. While on a high-protein diet or during starvation, humans synthesize urea cycle enzymes in amounts 5- to 10-fold higher than while on a low-protein diet.

Following removal of their amino groups, the carbon skeletons of all amino acids are degraded to intermediates already encountered in fuel metabolism: acetyl-CoA, pyruvate, and intermediates of the citric acid cycle (Fig 10–7). The reactions by which the carbon skeletons are converted to these intermediates will not be discussed in this text.

During fasting, those amino acids that are degraded to pyruvate or 4- or 5-carbon intermediates of the citric acid cycle can be used as substrates for gluconeogenesis. Amino acids that are degraded to acetyl-CoA provide substrates for ketogenesis. This distinction has led to the classification of amino acids as **glucogenic** or **ketogenic** (Table 10–2). Leucine and lysine are purely ketogenic; ie, their catabolism yields only acetyl-CoA. Four amino acids (isoleucine, phenylalanine, tryptophan, and tyrosine) are both glucogenic and ketogenic. The remaining 14 amino acids are glucogenic.

Genetic Defects of Amino Acid Metabolism

Derangements of amino acid metabolism are characterized by abnormal levels of amino acids or their metabolic products in the blood or urine. Defects in amino acid metabolism often cause mental retardation and developmental disorders. The precise nature of the metabolic defect has been identified in the case of several inborn errors of amino acid metabolism, including phenylketonuria, homocystinuria, and branched-chain ketonuria.

Table 10–2. Fates of the carbon skeletons of amino acids in fasting.

Glucogenic		Both Glucogenic and Ketogenic	Ketogenic
Ala	Arg	Ile	Leu
Asn	Asp	Phe	Lys
Cys	Glu	Trp	
Gln	Gly	Tyr	
His	Met		
Pro	Ser		
Thr	Val		

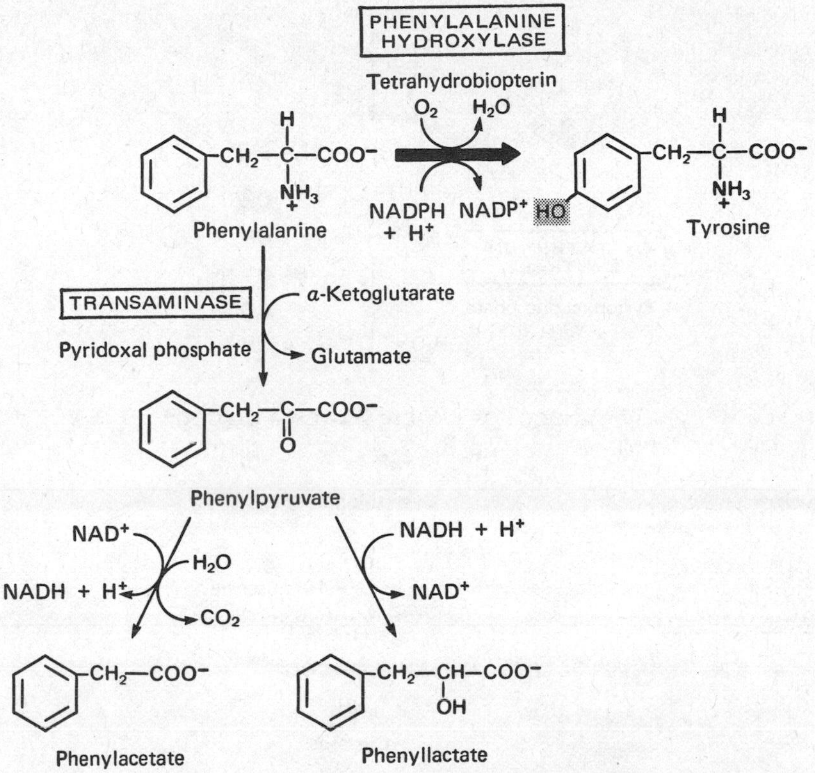

Figure 10–8. The major and minor pathways of phenylalanine metabolism.

In humans, the nonessential amino acid tyrosine is formed via hydroxylation of phenylalanine (Fig 10–8). This reaction is catalyzed by phenylalanine hydroxylase, which uses tetrahydrobiopterin, a compound resembling folic acid, as its coenzyme. Tetrahydrobiopterin serves as an intermediate carrier of reducing equivalents supplied ultimately by NADPH. Failure to carry out this reaction results in a disorder known as **phenylketonuria (PKU).** Because the reaction catalyzed by phenylalanine hydroxylase is both the only means of synthesizing tyrosine and the major means of catabolizing phenylalanine, in individuals with PKU, tyrosine is an essential amino acid and phenylalanine levels are greatly elevated. In normal individuals, a small amount of phenylalanine is converted to phenylpyruvate, phenylacetate, and phenyllactate. Because phenylalanine levels are elevated in individuals with PKU, these normally minor products are produced in greater amounts and are spilled in the urine. Some forms of PKU are due to genetic deficiency of phenylalanine hydroxylase; in others, production of tetrahydrobiopterin is defective. The major manifestation of the disease is mental retardation.

Methionine

```
          H
          |
-OOC—C—CH₂—CH₂—SH
          |
          NH₃
          +
```

$$\text{-OOC-C-CH}_2\text{-CH}_2\text{-SH}$$

Homocysteine

CYSTATHIONINE
SYNTHASE

Pyridoxal phosphate

OH

CH₂

—H—C—NH₃
       |
      COO⁻

Serine

H₂O

Cystathionine

```
          H                          H
          |                          |
-OOC—C—CH—CH₂-S—CH₂—C—COO⁻
          |                          |
          NH₃                        NH₃
          +                          +
```

Cystathionine

Homoserine

```
          H
          |
HS—CH₂—C—COO⁻
          |
          NH₃
          +
```

Cysteine

Figure 10–9. Homocystinuria is caused by a deficiency of cystathionine synthase.

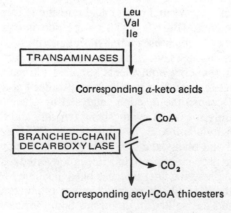

Leu
Val
Ile

TRANSAMINASES

Corresponding α-keto acids

CoA

BRANCHED-CHAIN
DECARBOXYLASE

CO₂

Corresponding acyl-CoA thioesters

Figure 10–10. Branched-chain ketonuria results from a defect in the pathway by which leucine, isoleucine, and valine are catabolized.

Deficiency of cystathionine synthase, an enzyme that forms part of the pathway for cysteine synthesis (Fig 10–9), results in a disorder known as **homocystinuria.** As the name suggests, this disorder is characterized by high urinary levels of homocysteine, a substrate of the impaired enzyme. Two forms have been described, one of which can be treated by high doses of vitamin B_6. This form of the disorder is due to the reduced affinity of cystathionine synthase for its coenzyme, pyridoxal phosphate. The other form is treated by limiting intake of methionine and by providing cysteine in the diet. The link between the biochemical lesion and its pathologic consequences is unknown.

Branched-chain ketonuria, also known as maple syrup urine disease, is the consequence of a defect in α-keto acid decarboxylase, an enzyme involved in the catabolism of leucine, valine, and isoleucine (Fig 10–10). The disease is manifested by vomiting, lethargy, and severe brain damage. Few infants survive beyond the first year of life. Again, the link between the observed derangement of metabolism and the symptoms of the disease is unknown.

11 | Heme Metabolism

OBJECTIVES

- Be able to describe the steps in the biosynthesis of heme. Know which step is rate-limiting and how the pathway is regulated.

- Be able to describe the biochemical consequences of a partial deficiency of each of the enzymes that participate in heme biosynthesis.

- Be able to describe the steps in the catabolism of heme and the biochemical consequences of deficiency of any of these reactions.

- Be able to describe how fetal bilirubin is excreted and to explain the basis of neonatal jaundice.

◄ • ►

HEME, THE PROSTHETIC group of hemoglobin and a number of other proteins, is a member of a group of colored compounds called porphyrins. A **porphyrin** is a cyclic compound whose structure includes 4 pyrrole rings and a system of conjugated double bonds (Fig 11–1). Heme is the most abundant porphyrin in human metabolism and is the only one for which there is a known function. The other porphyrins found in humans are by-products of heme biosynthesis.

Heme Biosynthesis

The pathway by which heme is synthesized is shown in Fig 11–2. It starts and ends inside the mitochondrion; however, 3 reactions take place in the cytoplasm. With the exception of mature erythrocytes, which lack

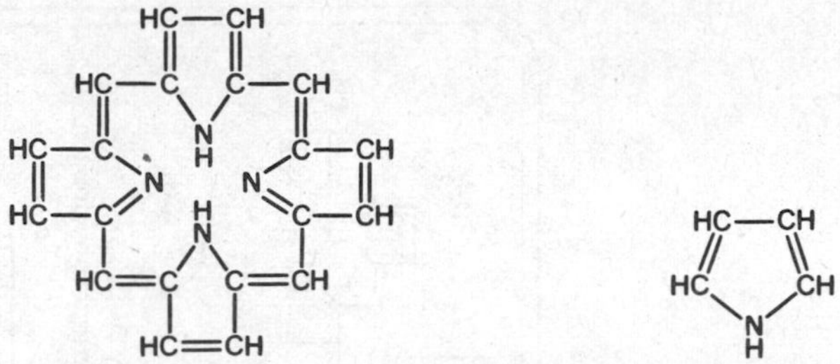

Figure 11–1. A porphyrin and a pyrrole.

mitochondria, all tissues produce some heme. The liver and bone marrow account for the largest amounts synthesized. Heme biosynthesis can be divided into 5 functional stages: (1) production of the monomeric pyrrole unit; (2) condensation of 4 pyrrole units to form a cyclic polymer; (3) modification of the side chains; (4) oxidation of the ring to form the system of conjugated double bonds; and (5) insertion of iron.

(1) The monomeric pyrrole unit of heme is constructed in 2 steps. First, glycine and succinyl-CoA condense to form δ-aminolevulinate. This reaction is catalyzed by the mitochondrial enzyme δ-aminolevulinate synthase. The product leaves the mitochondria by passive diffusion. In the cytoplasm, 2 molecules of δ-aminolevulinate combine to form porphobilinogen.

(2) Still in the cytoplasm, 4 porphobilinogen units condense to form a cyclic polymer, uroporphyrinogen. This and the subsequent 2 intermediates of the pathway lack the system of conjugated double bonds characteristic of porphyrins and are termed **porphyrinogens.** Because each of the pyrrole units of uroporphyrinogen has 2 different side chains (one acetate and one propionate), there are 4 possible isomers of uroporphyrinogen. In vivo, only 2 isomers are formed: the **type I isomer,** in which the pyrrole units are all condensed head to tail; and the **type III isomer,** which has one pyrrole unit in the opposite orientation. Heme is derived from the type III isomer. Formation of uroporphyrinogen is catalyzed by the enzyme uroporphyrinogen synthase, acting together with another protein, uroporphyrinogen cosynthase. Production of the asymmetric type III isomer depends on the activity of the cosynthase. Only when this protein is defective are substantial amounts of the type I isomer made.

(3) Modification of the side chains of uroporphyrinogen is performed in 2 steps, one cytoplasmic and one mitochondrial. First, the 4 acetate side chains are converted to methyl groups. The enzyme that catalyzes this step, uroporphyrinogen decarboxylase, acts on both the normal type III isomer and the abnormal type I isomer if it is present. However, only the type III product of the reaction, coproporphyrinogen III, is a substrate for the next

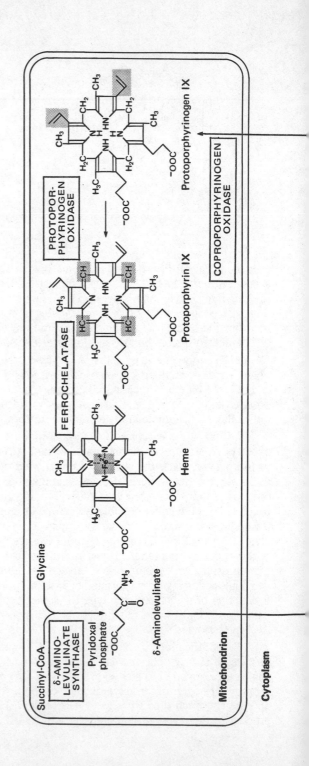

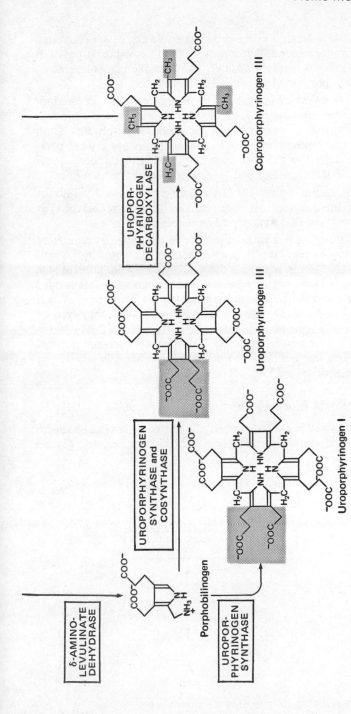

Figure 11–2. Synthesis of heme. (Redrawn, with permission, from Meyer UA, Schmid R: The porphyrias. In: *The Metabolic Basis of Inherited Disease*, 4th ed. Stanbury JB, Wyngaarden JB, Fredrickson DS [editors]. McGraw-Hill, 1978.)

enzyme of the pathway. Next, 2 of the propionate side chains are converted to vinyl groups, forming protoporphyrinogen. Of the 15 possible porphyrinogen isomers containing 4 methyl, 2 vinyl, and 2 propionate side chains, only one, designated protoporphyrinogen IX, is produced.

(4) Protoporphyrinogen IX is oxidized by the enzyme protoporphyrinogen oxidase, producing protoporphyrin IX. This oxidation step generates the system of conjugated double bonds characteristic of porphyrins. Both type I and type III uroporphyrinogens and coproporphyrinogens, when present, can also be oxidized to the corresponding porphyrins. Under normal circumstances, these intermediates and by-products do not accumulate.

(5) Heme biosynthesis is completed by insertion of ferrous iron into the porphyrin ring. If the enzyme that catalyzes this step, ferrochelatase, is defective, protoporphyrin IX spontaneously chelates zinc.

The activity of the heme biosynthetic pathway is controlled by regulating the synthesis of the first and rate-limiting enzyme of the pathway, δ-aminolevulinate synthase. Heme represses the synthesis of δ-aminolevulinate synthase (Fig 11–3). Thus, when heme is present in excess of that needed for the heme-binding apoproteins, synthesis of the enzyme is inhibited. When heme requirements increase, eg, if there has been substantial hemolysis or when synthesis of a hepatic heme-containing enzyme is induced, the synthesis of δ-aminolevulinate synthase is increased. The regulatory mechanism that mediates these effects has not yet been described (but see Chapter 13 for possible models of gene regulation).

Disorders of Porphyrin Biosynthesis

When the synthesis of heme is disturbed, the consequences can be severe for 2 reasons: (1) Underproduction of heme results in anemia, and (2) the

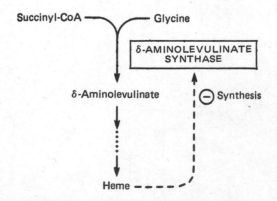

Figure 11–3. Control of heme biosynthesis.

accumulation of intermediates and by-products of the pathway is toxic. Disturbances of heme production, termed "porphyrias," can be either hereditary or acquired (caused by environmental poisons).

All porphyrias are characterized by accumulation of intermediates or by-products of the heme pathway, and the symptoms of each type can be understood in terms of the compounds that accumulate. Overproduction of δ-aminolevulinate and porphobilinogen is associated with abdominal pain,

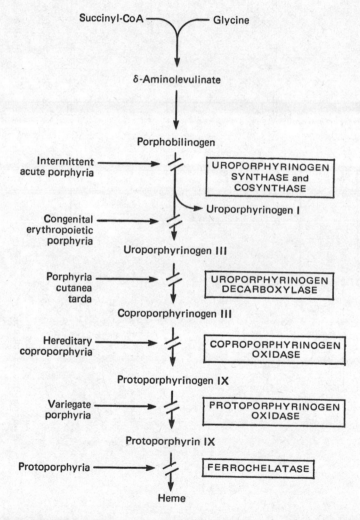

Figure 11–4. Hereditary porphyrias are caused by partial deficiencies of the enzymes required for heme biosynthesis.

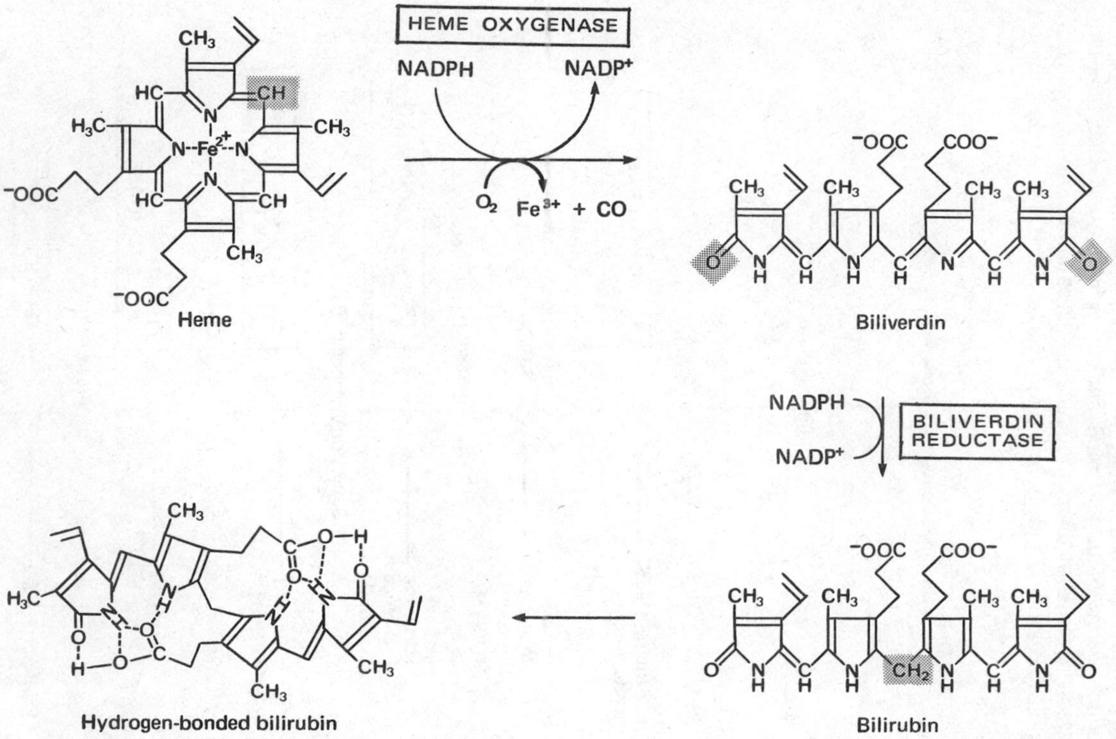

Figure 11–5. Heme is degraded by the heme oxygenase system.

vomiting, constipation, cardiovascular abnormalities, and neuropsychiatric signs, although the basis for these symptoms is obscure. An accumulation of by-product porphyrins (uroporphyrin I and III, coproporphyrin I and III, and protoporphyrin IX) results in a condition of photosensitivity that produces skin lesions. Owing to their system of conjugated double bonds, porphyrins absorb light. They may lose the additional energy contributed by the light by fluorescing or by transferring the energy to molecular oxygen, thereby producing an excited species called singlet oxygen. Singlet oxygen is highly reactive and oxidatively damages tissues in which it is produced. Heme also absorbs light but, owing to the presence of the iron atom, loses the additional energy without producing singlet oxygen.

Each of the hereditary porphyrias results from a partial deficiency of one of the enzymes of the pathway (Fig 11–4). Because heme mediates an essential function in humans, total absence of any enzyme would be lethal. Persons with **intermittent acute porphyria** are deficient in uroporphyrinogen synthase. These individuals accumulate δ-aminolevulinate and porphobilinogen. Deficiency of uroporphyrinogen cosynthase **(congenital erythropoietic porphyria)** results in overproduction of the type I porphyrins. The major symptom is photosensitivity. The remaining genetic porphyrias result from deficiencies in the pathway distal to the reaction catalyzed by uroporphyrinogen synthase and cosynthase. Each of these defects leads to an accumulation of all of the intermediates prior to the block and of the by-product porphyrins formed from these intermediates. Individuals so affected can therefore exhibit all of the possible symptoms of porphyria.

Heme Catabolism

The average life span of the red blood cell is 120 days. At the end of that time, it is removed from circulation by the spleen, liver, or bone marrow, and its component parts, including hemoglobin, are degraded. Heme-containing proteins of other tissues are turned over much more rapidly. In all cases, the proteins are hydrolyzed to free amino acids, and the heme ring is degraded. The Fe^{2+} is recovered for reuse. The first phase of heme catabolism yields a linear tetrapyrrole called **bilirubin,** which is only sparingly water-soluble and highly toxic to the nervous system. The remainder of the catabolic scheme is devoted to making bilirubin more soluble and excreting it from the body.

Heme catabolism begins with the oxidation of the heme ring by a heme oxygenase system associated with the endoplasmic reticulum of spleen, liver, and bone marrow (Fig 11–5). In this reaction (which requires O_2 and NADPH), the porphyrin is cleaved to form biliverdin and carbon monoxide and the iron atom is released. Biliverdin is rapidly reduced (via another NADPH-requiring reaction) to bilirubin. Although bilirubin contains 2 propionate side chains, it is largely insoluble in water because it is able to assume a conformation in which the side chains form intramolecular hydrogen bonds. Owing to its nonpolar character, bilirubin readily crosses cell membranes.

Bilirubin is transported in plasma by serum albumin. Under normal conditions, the concentration of bilirubin in circulation does not exceed the carrying capacity of albumin. However, the plasma concentration of bilirubin may be elevated by accelerated hemolysis or by a block in a subsequent step in heme catabolism. Furthermore, a number of drugs (eg, sulfonamides) compete for the bilirubin site on albumin and thus lower the capacity of plasma to carry bilirubin. Bilirubin in excess of that which can be carried by albumin readily leaves the circulatory system and enters extravascular tissues. Because bilirubin is yellow, this is seen as **jaundice** (also called icterus). Accumulation of bilirubin in the brain can result in brain damage (bilirubin encephalopathy).

Bilirubin is normally removed from circulation by the liver and made more polar by conjugation with one or 2 glucuronate residues (Fig 11–6). Glucuronate addition is catalyzed by bilirubin UDP-glucuronyltransferase. The product, called bilirubin glucuronide or **conjugated bilirubin,** is secreted by the liver into the bile. If bile secretion is blocked, conjugated bilirubin enters the circulation and is ultimately excreted in the urine. In the large intestine, bilirubin glucuronide secreted via the bile is acted upon by bacteria that remove the glucuronate residues and reduce the bilirubin to a series of compounds called urobilinogens. A fraction of the urobilinogen produced is reabsorbed by the intestines and resecreted in the bile. Ultimately, most of the urobilinogen is excreted from the body in the feces.

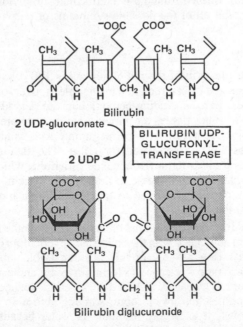

Figure 11–6. Conjugation of bilirubin with glucuronate.

Disorders of Heme Catabolism

A number of conditions result in abnormally high levels of plasma bilirubin (**hyperbilirubinemia**) in either the conjugated or the unconjugated form. Although both unconjugated and conjugated bilirubin can cause jaundice, only the unconjugated form is sufficiently nonpolar to cross the blood-brain barrier and cause brain damage. Unconjugated hyperbilirubinemia can result from (1) a transient neonatal deficiency of bilirubin UDP-glucuronyltransferase activity; (2) a genetic deficiency of the same enzyme; or (3) liver dysfunction due to, among other things, organic solvents, viral infection, or cirrhosis.

The human fetus relies on the maternal circulation for the excretion of bilirubin. The fetal liver contains very little glucuronyltransferase activity, and unconjugated bilirubin freely crosses the placenta to the maternal circulation. Synthesis of glucuronyltransferase is induced at birth and reaches full levels only after several months. Owing to immaturity of the conjugating system, many infants demonstrate a benign jaundice. However, bilirubin may reach toxic levels in infants slow to develop the conjugating system, especially premature infants, and in those in which hemolysis leads to the accelerated catabolism of heme.

Three medical syndromes are associated with genetic deficiencies of bilirubin UDP-glucuronyltransferase. The most severe is type I Crigler-Najjar syndrome, an autosomal recessive disorder in which the conjugating system is substantially defective. This syndrome is usually fatal within months of birth. In a milder form, type II Crigler-Najjar syndrome, the serum bilirubin levels usually do not exceed the capacity of serum albumin. A third genetic disorder, Gilbert's disease, is also associated with reduced glucuronyltransferase activity combined with hemolysis and a defect in the uptake of bilirubin into the liver.

Conjugated hyperbilirubinemia may result from either a genetic defect in the secretion of conjugated compounds from the liver (Dubin-Johnson syndrome) or obstruction of the hepatic or common bile ducts. In both cases, conjugated bilirubin enters the plasma and is excreted in the urine.

12 | Metabolism of Purine & Pyrimidine Nucleotides

OBJECTIVES

- Learn the system of nomenclature used to describe nucleotides and their component parts. Know the names of the common purine and pyrimidine bases, nucleosides, and nucleotides.

- Be able to name the substrates and products of the pathways involved in the de novo synthesis of ribonucleotides. Know the origin of each of the atoms of the purine and pyrimidine bases. Be able to describe the means by which these pathways are regulated.

- Be able to name the enzyme that catalyzes the formation of deoxy-ribonucleotides, to name its substrates, and to describe its regulation.

- Be able to describe the synthesis of the thymine-containing nucleotides.

- Understand how vitamin B_{12} and the folate coenzymes participate in nucleotide metabolism, and be able to explain the pathologic effects of deficiency of these coenzymes. Know the biochemical basis for the therapeutic use of sulfonamides and dihydrofolate reductase inhibitors.

- Be able to explain how purine nucleotides are catabolized and the relationship of purine catabolism to hyperuricemia and gout.

- Be able to describe the roles of hypoxanthine-guanine phosphoribosyl transferase, purine nucleoside phosphorylase, and adenosine deaminase in the salvage of purine nucleotides. Be able to explain the biochemical consequences of genetic deficiencies of each enzyme.

◄ • ►

PURINE AND pyrimidine nucleotides are small nitrogen-containing compounds that play many important biologic roles. Among other things, nucleotides serve as carriers of metabolic energy (eg, ATP), as substrates for the synthesis of RNA and DNA, as components of coenzymes (eg, NAD), and as allosteric regulators of enzymatic activity. As illustrated in Fig 12–1, a **nucleotide** consists of a purine or pyrimidine base joined by a glycosidic linkage to a pentose sugar, which in turn is esterified to one or more phosphate groups. Note that the atoms of the bases and sugars are numbered independently and that the sugar carbons are designated 1′ through 5′. Included within the structure of the nucleotide is a **nucleoside,** ie, a base joined to a pentose. For this reason, nucleotides are also referred to as phosphorylated nucleosides. For example, a purine nucleotide that has one phosphate group attached to the 5′ carbon is termed a purine nucleoside 5′-monophosphate.

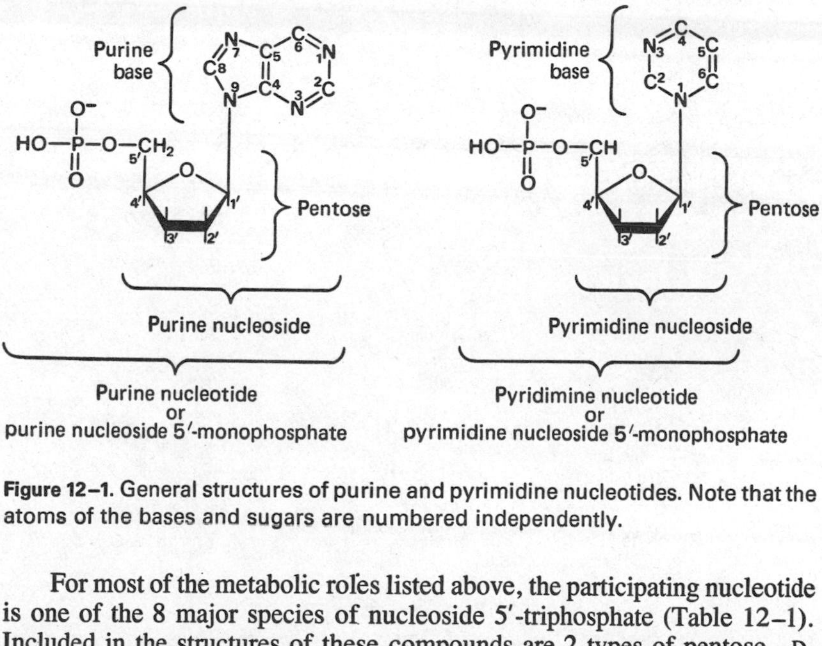

Figure 12–1. General structures of purine and pyrimidine nucleotides. Note that the atoms of the bases and sugars are numbered independently.

For most of the metabolic roles listed above, the participating nucleotide is one of the 8 major species of nucleoside 5′-triphosphate (Table 12–1). Included in the structures of these compounds are 2 types of pentose—D-ribose and 2-deoxy-D-ribose—and 5 types of base—adenine, guanine, cytosine, uracil, and thymine (Fig 12–2).

Nucleotide metabolism involves several interconnected pathways. Ribonucleotides are formed both de novo, using products of other metabolic pathways, and by salvage of nucleotides released in the degradation of nucleic acids. The deoxyribonucleotides are synthesized from ribonucleotides by reduction of the pentose moiety. Nucleotides in excess of those needed for anabolism are degraded to products that can either be consumed by other

Table 12-1. Major species of nucleoside triphosphates.

Nucleoside Triphosphates	Names of Components	
	Bases	Nucleosides
Ribonucleotides		
Adenosine triphosphate (ATP)	Adenine	Adenosine
Guanosine triphosphate (GTP)	Guanine	Guanosine
Cytidine triphosphate (CTP)	Cytosine	Cytidine
Uridine triphosphate (UTP)	Uracil	Uridine
Deoxyribonucleotides		
Deoxyadenosine triphosphate (dATP)	Adenine	Deoxyadenosine
Deoxyguanosine triphosphate (dGTP)	Guanine	Deoxyguanosine
Deoxycytidine triphosphate (dCTP)	Cytosine	Deoxycytidine
Deoxythymidine triphosphate (dTTP)	Thymine	Deoxythymidine

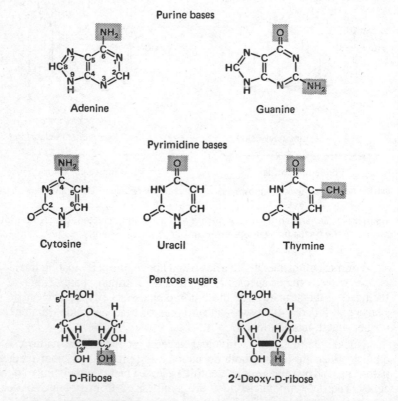

Figure 12-2. Components of the common nucleotides. Shading indicates the distinctive features of each compound.

pathways or excreted. This chapter outlines each of these processes, with an emphasis on the substantial interdependence and overlap of the pathways.

Folate Coenzymes & Vitamin B_{12}

Metabolites of 2 water-soluble vitamins, folic acid and vitamin B_{12} (cobalamin), play important roles in nucleotide metabolism. In the absence of either of these vitamins, nucleotide biosynthesis is inhibited. This in turn

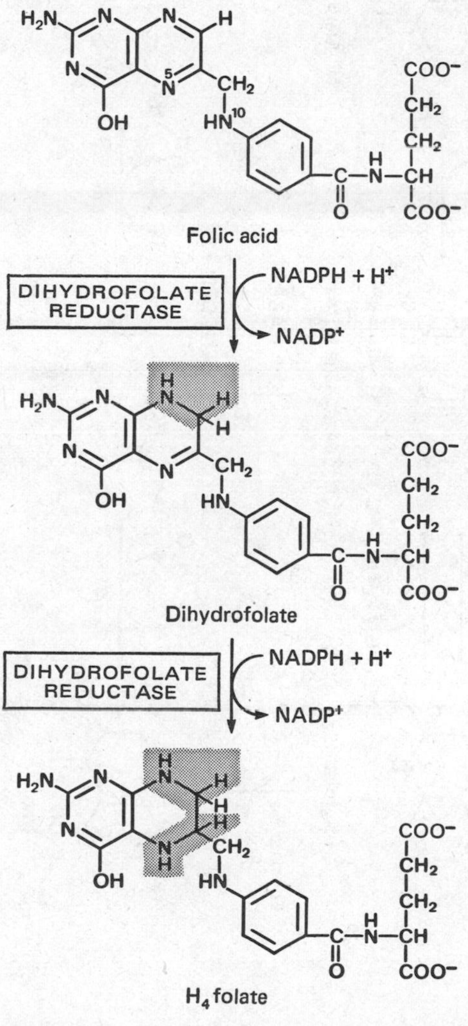

Figure 12–3. Folic acid is reduced in 2 steps to H_4folate.

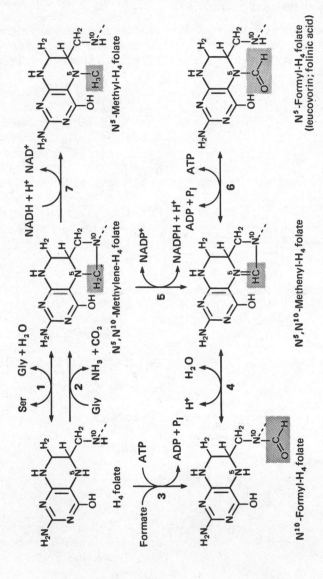

Figure 12–4. Formation of the one-carbon derivatives of H₄folate. The dashed line indicates that only a portion of the structure is shown.

prevents the growth and development of many tissues and in particular causes a megaloblastic form of anemia (*megalo* = large, *blast* = immature stage in cellular development).

* Folic acid functions metabolically as a carrier of one-carbon units; ie, it mediates the transfer from one compound to another of functional groups that contain one carbon atom. In order to participate in one-carbon transfer, folic acid must first be reduced to **tetrahydrofolate (H$_4$folate)**. Reduction is carried out in 2 steps by dihydrofolate reductase, using NADPH as the source of reducing equivalents (Fig 12–3).

Fig 12–4 illustrates the synthesis of several of the one-carbon derivatives of H$_4$folate. N^5,N^{10}-Methylene-H$_4$folate is formed by transfer of a methylene group from either serine or glycine to H$_4$folate (reactions 1 and 2). Condensation of H$_4$folate with formate yields N^{10}-formyl-H$_4$folate (reaction 3). These folate derivatives can be converted to the others, as shown in the figure. Note that while most of the interconversions are reversible, the reduction of N^5,N^{10}-methylene-H$_4$folate to form N^5-methyl-H$_4$folate is not (see reaction 4). Most of the folates consumed in food are converted to N^5-methyl-H$_4$folate during their absorption from the intestine.

The methylene, methenyl, and N^{10}-formyl derivatives of H$_4$folate participate in the synthesis of nucleotides (see below). N^5-Methyl-H$_4$folate donates its methyl group to homocysteine to form methionine (Fig 12–5). The enzyme that catalyzes this transfer, homocysteine methyltransferase, uses cobalamin as its prosthetic group. In this reaction, cobalamin functions as an intermediate carrier of the methyl group that is being transferred. Note that this transfer provides a mechanism for biosynthesis of methionine, a nutritionally essential amino acid. Despite the operation of this reaction, there is no net synthesis of methionine in humans, because the only route by which homocysteine can be synthesized is through the degradation of methionine (Fig 12–5). The reaction catalyzed by homocysteine methyltransferase is important primarily because it converts methyl-H$_4$folate to H$_4$folate, which can then react with serine, glycine, or formate to form the other folate derivatives. Because most dietary folates are converted to methyl-H$_4$folate during absorption and because methyl-H$_4$folate can be converted to the other folate derivatives only via the reaction involving vitamin B$_{12}$, a deficiency of vitamin B$_{12}$ traps the bulk of the folate pool as methyl-H$_4$folate and thus causes a deficiency of the other folate derivatives.

Another compound active in the transfer of one-carbon units is S-adenosylmethionine (SAM). SAM is formed by the condensation of methionine with the adenosine moiety of ATP (Fig 12–5). It serves as the methyl group donor in the synthesis of a number of compounds such as phosphocreatine, a high-energy phosphate compound active in muscle ATP production, and epinephrine, a hormone.

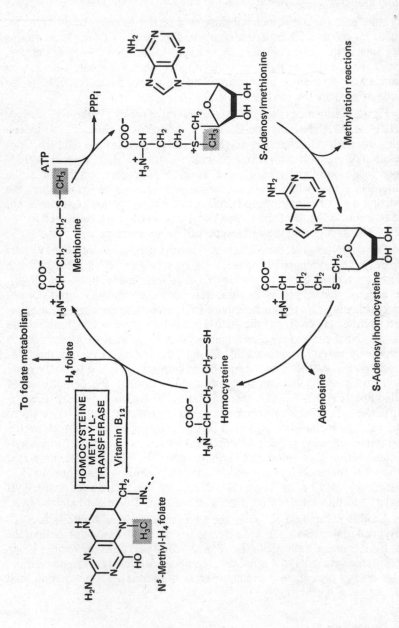

Figure 12–5. Transfer of the methyl group of N^5-methyl-H_4folate to homocysteine is catalyzed by a cobalamin-dependent enzyme.

De Novo Biosynthesis of Ribonucleotides

The purine and pyrimidine nucleotides are synthesized de novo from amino acids, CO_2, one-carbon groups carried by the folate cofactors, and ribose 5-phosphate provided by the pentose phosphate pathway (see Chapter 9). Thus, the raw materials for both types of nucleotide have a common origin. However, the pathways by which they are formed are distinct in design. In the pyrimidine pathway, the ring structure of the base is first assembled and then attached to the pentose sugar. In contrast, the purine pathway starts with the pentose sugar and builds the ring structure of the base upon it. In both pathways, the immediate precursor of the sugar moiety is 5-phosphoribosyl-1-pyrophosphate (PRPP). This activated derivative of ribose is formed from ribose 5-phosphate and ATP (Fig 12–6). PRPP is consumed by a number of metabolic pathways, including de novo nucleotide synthesis, salvage of nucleotides, and synthesis of nucleotide coenzymes (eg, NAD).

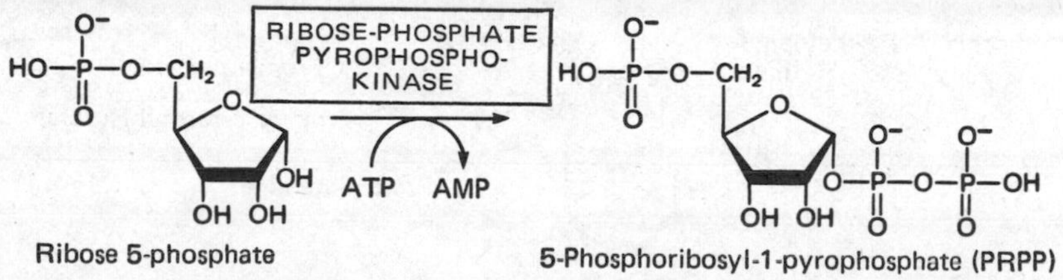

Figure 12–6. Synthesis of PRPP.

The pathway for the de novo synthesis of purine ribonucleotides is shown in abbreviated form in Fig 12–7. In the first and rate-limiting reaction, glutamine donates its amide group to carbon 1 of PRPP, thus attaching to the sugar what will become nitrogen 9 of the completed purine ring. This constitutes the committed step in purine biosynthesis. The remaining atoms of the purine ring are then added stepwise. Carbons 4 and 5 and nitrogen 7 are derived from glycine, nitrogen 3 from glutamine, and nitrogen 1 from aspartate. Carbon 6 is added in a biotin-independent carboxylation reaction. Carbons 2 and 8 are donated by folate cofactors N^{10}-formyl-H_4folate and N^5,N^{10}-methenyl-H_4folate, respectively. Because the ribose and phosphate moieties are incorporated at the beginning of the pathway, completion of the purine base results in the formation of a purine nucleoside monophosphate, inosine 5′-monophosphate (IMP), rather than a free purine base. Both adenosine monophosphate (AMP) and guanosine monophosphate (GMP) are formed from IMP.

Nucleoside 5′-monophosphates are phosphorylated to the corresponding diphosphates by a group of nucleoside monophosphate kinases each, of which is specific for the base component of the nucleotide. Nucleoside diphosphates and triphosphates are interconverted by a nucleoside diphosphate kinase that

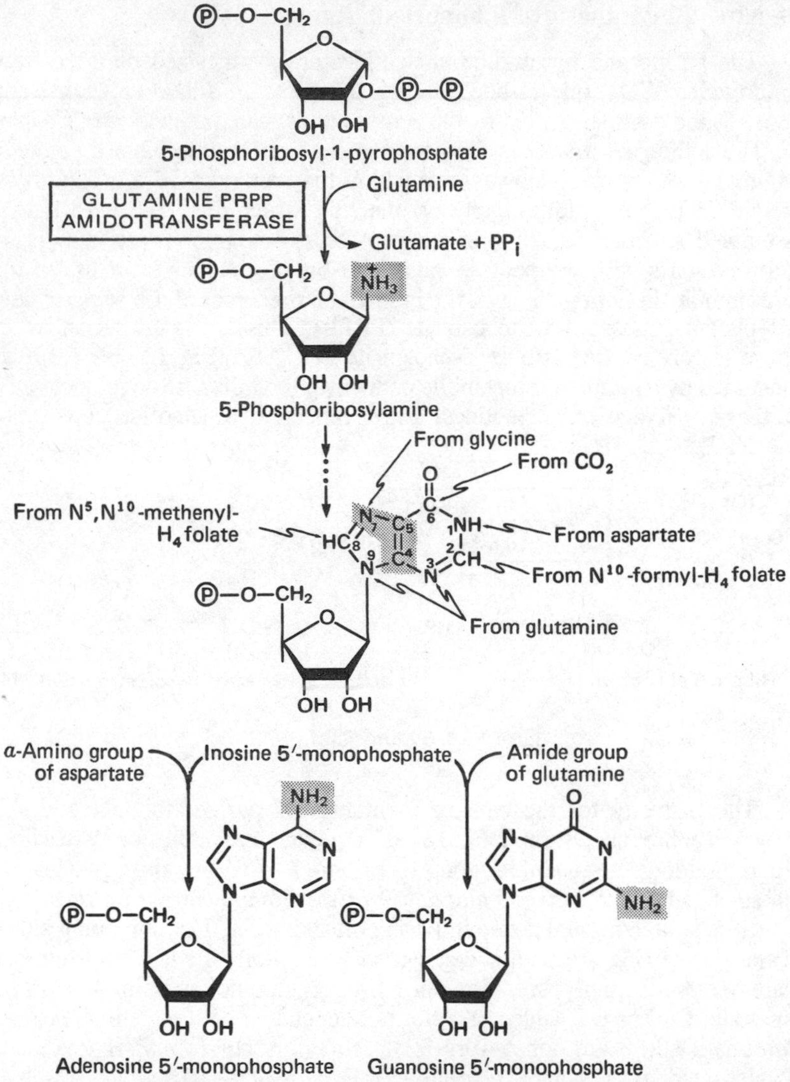

Figure 12–7. De novo synthesis of the purine nucleotides. The notation $\rightarrow \cdots \rightarrow$ indicates that several intervening steps are not shown. $\textcircled{P}$ = phosphoryl group.

does not have a requirement for a specific base. ATP is the usual phosphate donor in all of these phosphorylations.

De novo pyrimidine biosynthesis begins with the formation of carbamoyl phosphate from the amide group of glutamine, CO_2, and a phosphoryl group of ATP (Fig 12–8). Carbamoyl phosphate is both an intermediate in pyrim-

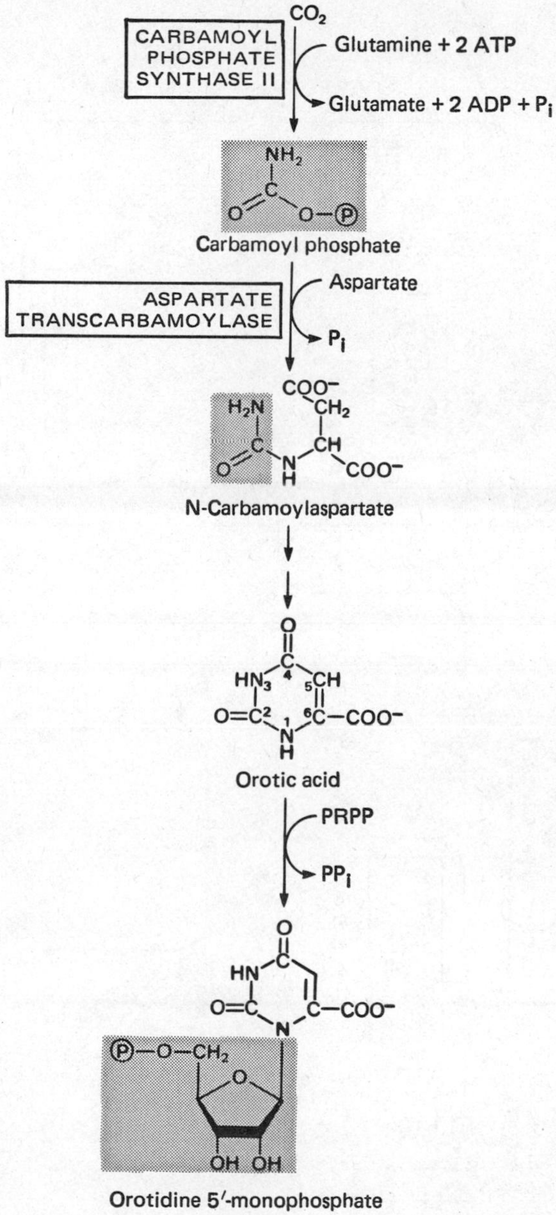

Figure 12–8. De novo synthesis of pyrimidine nucleotides. $\textcircled{P}$ = phosphoryl group.

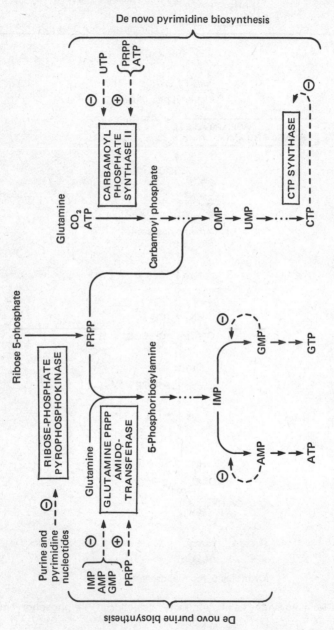

Figure 12–9. Regulation of the de novo synthesis of nucleotides.

idine biosynthesis and a substrate of the urea cycle. Although both pathways utilize the same compound, they do not interact, because they are located in different compartments of the cell. Carbamoyl phosphate destined for the urea cycle is synthesized in the mitochondria in a reaction catalyzed by carbamoyl phosphate synthase I (see Chapter 10). The enzyme that catalyzes the first reaction of the pyrimidine pathway, carbamoyl phosphate synthase II, is a component of cytoplasm.

Carbamoyl phosphate is the origin of carbon 2 and nitrogen 3 of the pyrimidine ring. The remaining atoms of the ring are added as a unit in the form of aspartate. The resulting N-carbamoyl aspartate is converted to a free pyrimidine base, orotate, by ring closure and oxidation. The base is then joined to PRPP to form a nucleoside monophosphate, orotidine monophosphate (OMP). Uridine monophosphate (UMP) is derived directly from OMP by decarboxylation. UMP is phosphorylated to form uridine triphosphate (UTP), which accepts the amide group of glutamine to become cytidine triphosphate (CTP).

Control of Ribonucleotide Biosynthesis

The pathways for the de novo synthesis of purines and pyrimidines are primarily regulated by the concentrations of their own products. This form of regulation ensures an adequate supply of nucleotides while preventing their overproduction. Both pathways are also affected by the concentration of PRPP. The magnitude of the PRPP pool is primarily affected by the rate at which it is synthesized via ribose-phosphate pyrophosphokinase and by the rate at which it is utilized by the salvage pathways (see below). In normal individuals, ribose-phosphate pyrophosphokinase is feedback-inhibited by both purine and pyrimidine nucleotides (Fig 12–9). Individuals with abnormal forms of ribose-phosphate pyrophosphokinase, either superactive or resistant to feedback inhibition, have been identified.

The first and rate-limiting enzyme of the purine pathway, glutamine PRPP amidotransferase, is feedback-inhibited by a number of purine nucleotides, including IMP, AMP, ADP, GMP, and GDP (Fig 12–9). If, however, the intracellular concentration of PRPP rises above normal levels, glutamine PRPP amidotransferase escapes from feedback inhibition, and regulation of the purine pathway is impaired. Thus, individuals with superactive or feedback-resistant forms of ribose-phosphate pyrophosphokinase overproduce the purine nucleotides. Additional controls on the purine pathway regulate the production of AMP and GMP from IMP. The adenine and guanine nucleotides each inhibit their own production.

The first enzyme of the pyrimidine pathway, carbamoyl phosphate synthase II, is feedback-inhibited by UTP and stimulated by ATP. Thus, as ATP levels rise, they stimulate a corresponding increase in pyrimidine production. CTP synthase is feedback-inhibited by its product, CTP.

Deoxyribonucleotide Biosynthesis

The deoxyribonucleotides needed for DNA synthesis are formed from ribonucleotides by the reduction of the pentose ring at the 2' position. A single enzyme, ribonucleotide reductase, catalyzes the conversion of each of the ribonucleotide *di*phosphates to the corresponding deoxyribonucleoside *di*phosphates (Fig 12–10). NADPH donates the reducing equivalents used in this reaction via an unknown intermediate.

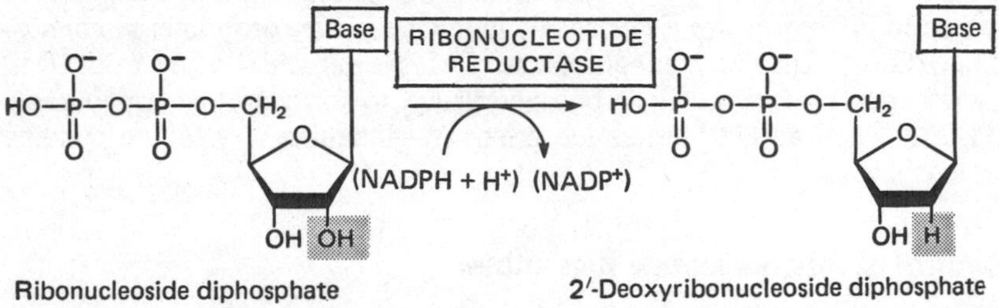

Ribonucleoside diphosphate   2'-Deoxyribonucleoside diphosphate

Figure 12–10. Synthesis of the deoxyribonucleotides is catalyzed by ribonucleotide reductase. Reducing equivalents used in this reaction are provided by NADPH and are transferred to the ribonucleotide via an unidentified carrier.

The activity of ribonucleotide reductase is regulated so as to ensure that deoxyribonucleotide synthesis is balanced, ie, that none of the deoxyribonucleotides are synthesized in amounts disproportionate to the others. This type of regulation is important because an imbalance in the deoxyribonucleotide pools would decrease the accuracy of DNA synthesis and consequently increase the rate of mutation (see Chapter 13). Ribonucleotide reductase is allosterically regulated by ATP, dTTP, dGTP, and dATP (Fig 12–11). ATP increases the rate at which the enzyme converts cytosine diphosphate (CDP) and uridine diphosphate (UDP) to the corresponding deoxyribonucleotides. DeoxyUDP is converted, in several steps, to dTTP (see below). The latter inhibits ribonucleotide reductase with respect to CDP and UDP but activates it with respect to guanosine diphosphate (GDP) and adenosine diphosphate (ADP). A pyrimidine deoxyribonucleotide thus feedback-inhibits its own production and stimulates the synthesis of the purines. Similarly, dGTP inhibits its own production and that of the pyrimidines while stimulating the production of dATP. Finally, dATP inhibits the enzyme with respect to all 4 ribonucleotide substrates and thus shuts off deoxyribonucleotide synthesis. If, as a result of overactive salvage pathways, excess dATP or dGTP accumulates, it prevents the synthesis of the pyrimidine deoxyribonucleotides, with the result that DNA synthesis is inhibited.

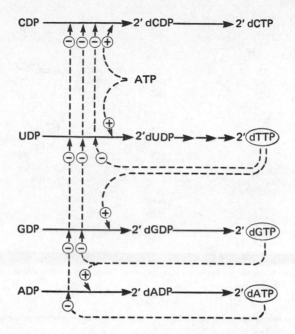

Figure 12–11. Complex regulation of ribonucleotide reductase ensures the production of balanced pools of deoxyribonucleotides. Solid lines represent chemical flow, and dotted lines represent negative ($\ominus$) or positive ($\oplus$) feedback regulation. (Reproduced, with permission, from Martin DW Jr et al: *Harper's Review of Biochemistry,* 20th ed. Lange, 1985.)

The thymine-containing nucleotides arise via a reaction in which deoxyuridine monophosphate (dUMP) is methylated to form deoxythymidine monophosphate (dTMP) (Fig 12–12). Thymidylate synthase catalyzes both the transfer of the methylene group of N^5, N^{10}-methylene-H_4folate to dUMP and the simultaneous reduction of the transferred group by the folate cofactor. This reaction is unusual among those in which the folates participate, because the cofactor is oxidized to H_2folate in the process. Before it can again become involved in one-carbon transfer, H_2folate must be reduced by dihydrofolate reductase.

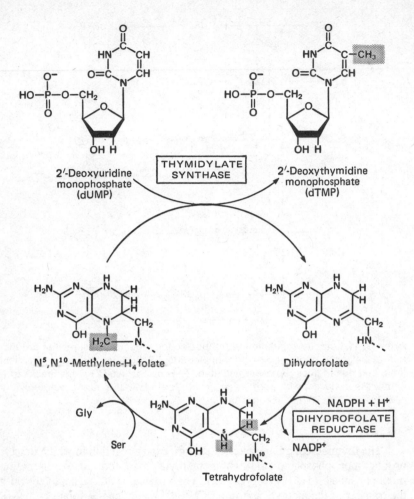

Figure 12-12. Thymidylate synthase catalyzes the synthesis of dTMP from dUMP.

Inhibitors of Tetrahydrofolate Metabolism

Owing to their central role in nucleotide metabolism, the folate cofactors are necessary for the growth of all known organisms. Drugs that block the formation of H₄folate ("antifolates") are effective in the treatment of bacterial infections and some forms of cancer.

While bacteria are able to synthesize folic acid de novo, humans cannot and must therefore obtain it from the diet. This metabolic difference accounts for the selective toxicity of the sulfonamide antibiotics toward bacteria. The sulfonamides (eg, sulfanilamide) are analogs of *p*-aminobenzoic acid (PABA),

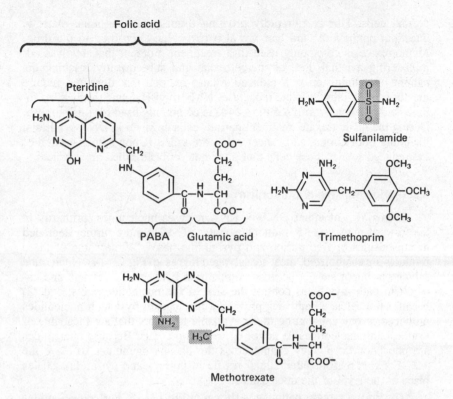

Figure 12–13. Analogs of *p*-aminobenzoic acid and folate are used therapeutically.

a component of folic acid (Fig 12–13). When exposed to a sulfonamide, bacteria use it instead of PABA. The resulting folate analog is inactive in one-carbon transfer and thus inhibits nucleotide synthesis in the affected bacteria. Because humans do not have the enzymes needed for the synthesis of folic acid from PABA, folate metabolism in humans is unaffected by sulfonamides.

Inhibitors of dihydrofolate reductase also have therapeutic applications. Trimethoprim (Fig 12–13) is used as an antibiotic because it strongly inhibits bacterial dihydrofolate reductase at concentrations that have little effect on the corresponding human enzyme. Another inhibitor of dihydrofolate reductase, methotrexate, inhibits the bacterial and human enzymes equally and therefore cannot be used as an antibiotic. It does, however, have a role in cancer chemotherapy. Cells that are actively engaged in DNA synthesis are particularly sensitive to methotrexate toxicity owing to the oxidation of H$_4$folate to H$_2$folate that accompanies synthesis of dTMP. Because cancer cells are engaged in DNA synthesis for a larger fraction of their cell cycle than are normal cells, methotrexate treatment kills more cancer cells than normal cells. However, rapidly growing tissues, such as bone marrow, intestinal epithelium, and hair are also particularly sensitive to the drug. Methotrexate is commonly used in a treatment protocol that involves administering a lethal dose of methotrexate and subsequently rescuing the patient by administering a reduced folate, ie, one that does not require metabolism by dihydrofolate reductase. N^5-Formyl-H_4folate (also called folinic acid and leucovorin; see Fig 12–4) is commonly used for this purpose. During the methotrexate treatment period, cells engaged in DNA synthesis, including both normal and cancer cells, are killed. Upon administration of the reduced folate, those cells that have not yet been killed are rescued.

Nucleotide Salvage & Catabolism

Breakdown of either DNA or RNA releases nucleotides, primarily in the form of nucleoside 5'-monophosphates. The latter are further degraded to nucleosides or free purine and pyrimidine bases. While some of these products are catabolized, most are salvaged for reuse (Fig 12–14). Nucleotide salvage is important (1) because competition between the salvage and catabolism pathways helps control the size of the nucleotide pools; and (2) because the salvage pathways provide a mechanism by which nucleotides produced in one cell can be made available to others that are incapable of synthesizing nucleotides de novo, eg, red blood cells. Because nucleotides are phosphorylated, they cannot cross the plasma membrane to leave the cell. For this reason, intercellular traffic in purines and pyrimidines takes place at the level of the nucleosides.

The purine salvage pathway is shown in Fig 12–15. Purine nucleotides are first hydrolyzed to the corresponding nucleosides by 5'-nucleotidase. The nucleosides may be salvaged in the same cell or may enter the circulation and be transported to other cells. Some purine nucleosides are salvaged

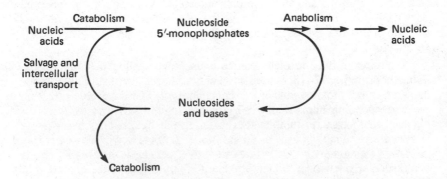

Figure 12–14. Nucleotides released in the breakdown of nucleic acids may be catabolized or salvaged.

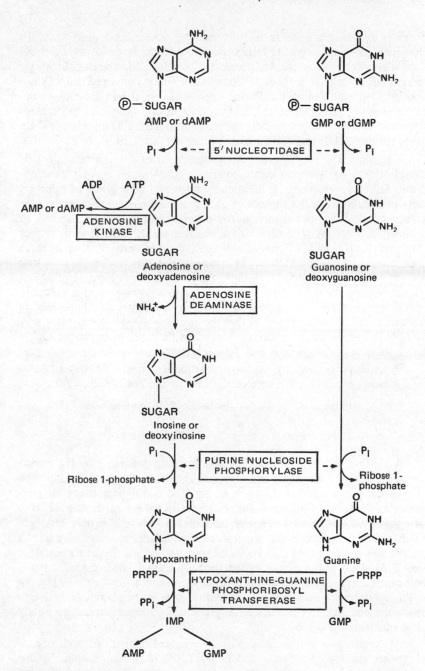

Figure 12–15. Salvage of the purine nucleotides. ⓟ = phosphoryl group; – – ⟶ = catalysis.

directly; eg, most adenosine is converted by adenosine kinase to AMP. However, most are degraded to free purine bases prior to salvage. Deoxyadenosine, deoxyguanosine, and guanosine largely follow the latter pathway. In a normal individual, most of the deoxyadenosine is converted to deoxyinosine by adenosine deaminase. Purine nucleoside phosphorylase cleaves deoxyinosine to hypoxanthine and deoxyguanosine and guanosine to guanine. Hypoxanthine-guanine phosphoribosyl transferase (HGPRTase) salvages these bases to form GMP and IMP, which reenter the nucleotide pools.

A genetic deficiency of either adenosine deaminase or purine nucleoside phosphorylase has pathologic consequences. Individuals lacking these enzymes fail to develop normal immune systems and usually die of infection early in childhood. In the absence of adenosine deaminase, deoxyadenosine does not follow its normal salvage route, which leads to IMP, but instead is phosphorylated by adenosine kinase to form dAMP. The latter is further phosphorylated to dATP. Use of this alternative pathway results in an increase in the concentration of dATP, which in turn inhibits ribonucleotide reductase and prevents DNA synthesis. A deficiency of purine nucleotide phosphorylase similarly shunts deoxyguanosine through a pathway that is normally minor and results in levels of dGTP that are inhibitory to ribonucleotide reductase. It has been suggested that the immune system is uniquely susceptible to imbalance of the deoxyribonucleotide pools owing to the highly active salvage pathways in B and T lymphocytes.

Pyrimidine nucleosides are salvaged directly. Each is phosphorylated by a kinase specific for the base component of the nucleoside, eg,

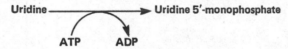

A portion of the hypoxanthine and guanine produced by the purine salvage pathway is degraded as shown in Fig 12–16. Through the activities of guanase and xanthine oxidase, both guanine and hypoxanthine are converted to xanthine. Xanthine is further metabolized by xanthine oxidase to **uric acid,** the end product of purine catabolism. Uric acid is only sparingly soluble and is excreted in the urine, diluted with large amounts of water. If the plasma level of uric acid becomes unusually high (hyperuricemia), it may precipitate in the tissues in the form of sodium urate crystals. Upon ingesting the crystals, macrophages initiate an inflammatory response leading to the syndrome known as **gout.** Hyperuricemia is, in many cases, due to an accelerated production of uric acid, secondary to degradation of unusually large quantities of purines.

Hyperuricemia is associated with genetic deficiencies of both ribose-phosphate pyrophosphokinase and HGPRTase. If ribose-phosphate pyrophosphokinase is overactive or is resistant to feedback inhibition, PRPP is overproduced. The increased concentration of PRPP causes glutamine PRPP amidotransferase, the rate-limiting enzyme in purine biosynthesis, to escape

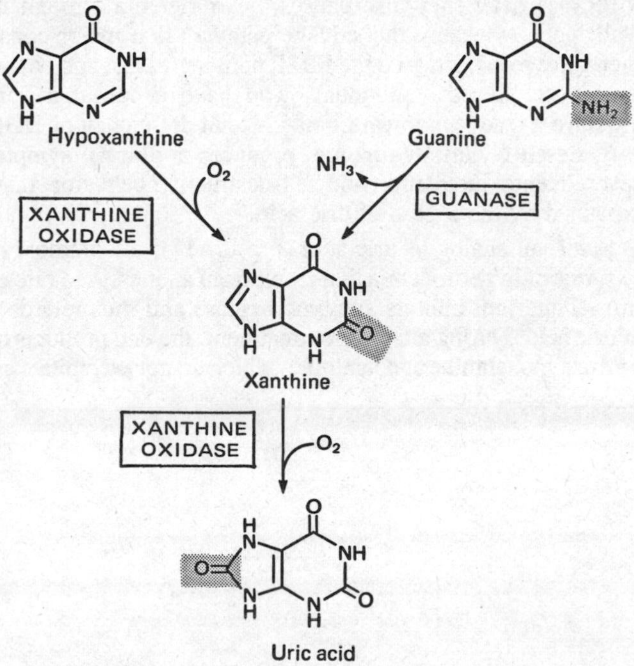

Figure 12–16. Uric acid is the end product of purine catabolism.

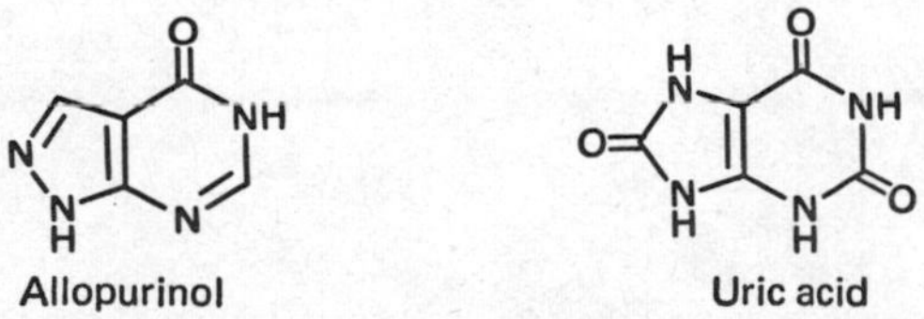

Figure 12–17. Allopurinol, an analog of uric acid, is used in the treatment of hyperuricemia.

from end-product inhibition, with the result that purines are overproduced. Overproduction in turn leads to increased catabolism and elevated levels of urate. HGPRTase deficiency also causes hyperuricemia through its effect on the PRPP pools. Because the salvage pathway is a major consumer of PRPP, when salvage is decreased the PRPP pool increases, and consequently purines are overproduced. Individuals who have a partial deficiency of HGPRTase have a tendency toward gout. A total deficiency of HGPRTase, the cause of **Lesch-Nyhan syndrome,** produces additional symptoms, including severe mental retardation and self-destructive behavior, that are not readily explained by the excess of uric acid.

Allopurinol, an analog of uric acid (Fig 12–17), is commonly used to treat those forms of hyperuricemia that result from an increased rate of purine catabolism. Allopurinol inhibits xanthine oxidase and thus retards the production of uric acid. During allopurinol treatment, the end products of purine catabolism are hypoxanthine and xanthine, which do not precipitate as readily as urate.

Section III: Cell Biology

Genes & Gene Expression | 13

OBJECTIVES

- Be able to describe the structures of DNA and RNA and to explain how these macromolecules encode the structures of proteins.

- Be able to describe the structure and the features of the genetic code and to explain the effects of substitution, insertion, and deletion mutations on the function of the code.

- Be able to describe the histones and the role they play in chromatin structure.

- Be able to explain how DNA is replicated and repaired.

- Be able to describe the steps in the synthesis of RNA and protein and to explain how these processes differ in prokaryotes and eukaryotes.

- Be able to describe the operon model for regulation of bacterial transcription and to predict how a mutation in each of the elements of this system would affect transcription.

BECAUSE PROTEINS catalyze most of the biologically important reactions, every organism must have the ability to make these macromolecules. The information needed for the synthesis of proteins—information for the primary structures of the proteins themselves and for molecules that participate in protein synthesis—is encoded in genetic material. This chapter describes the structure of genetic material, how it is maintained and duplicated, how it codes for proteins, and how proteins are synthesized.

Earlier chapters in this book focused on the biochemistry of eukaryotic cells, ie, cells that are divided by internal membranes into subcellular compartments such as the nucleus, mitochondria, and endoplasmic reticulum. **Eukaryotes** were emphasized because human cells are eukaryotic. In this chapter, some time will be devoted to the discussion of bacteria, which are **prokaryotes.** The prokaryotic cell is not subdivided by internal membranes. While the fundamental mechanisms by which genetic material is maintained and expressed are similar in eukaryotes and prokaryotes, the differences that have been observed are interesting from an evolutionary point of view and also provide the basis for some types of antibiotic therapy.

DNA Structure

Most forms of life, including all single- and multi-cell organisms and many viruses, use **DNA (deoxyribonucleic acid)** as their genetic material. DNA is a very long, unbranched polymer the monomeric units of which are

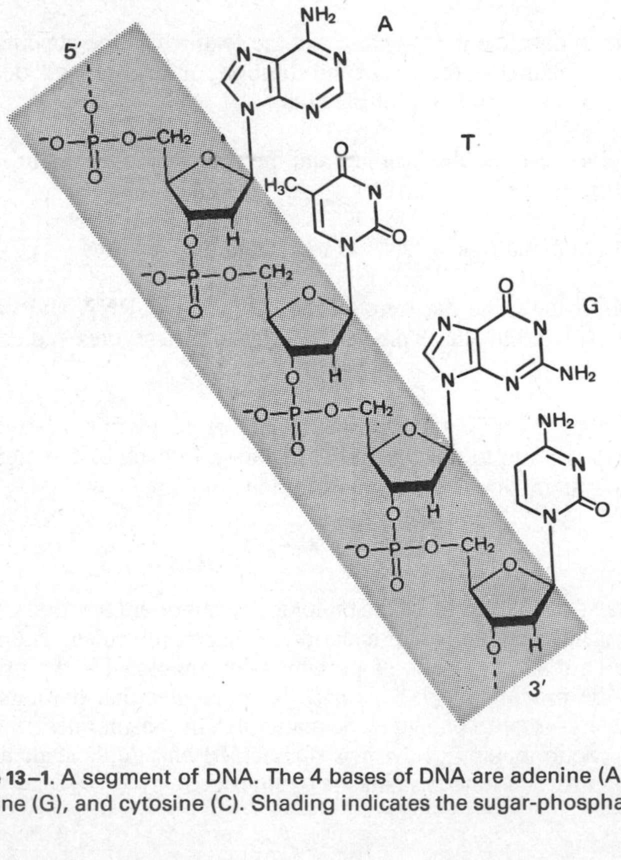

Figure 13–1. A segment of DNA. The 4 bases of DNA are adenine (A), thymine (T), guanine (G), and cytosine (C). Shading indicates the sugar-phosphate backbone.

4 deoxyribonucleoside monophosphates: dAMP, dGMP, dCMP, and dTMP (see Fig 13–1 and Chapter 12). In DNA, these units are joined to one another by ester bonds that link the 3′ hydroxyl of one nucleotide to the 5′ phosphate of the next. The resulting polymer has a repetitive backbone of alternating sugar and phosphate groups and a variable sequence of bases. Because each phosphate group in the backbone is joined by ester linkages to 2 sugar units, DNA is said to have a phosphodiester backbone.

Each DNA chain has a **polarity;** ie, the 2 ends of the polymer are different. At one end of the DNA (the 5′ end), the carbon atom nearest to the end of the polymer is the 5′ carbon of the last deoxyribose unit. At the

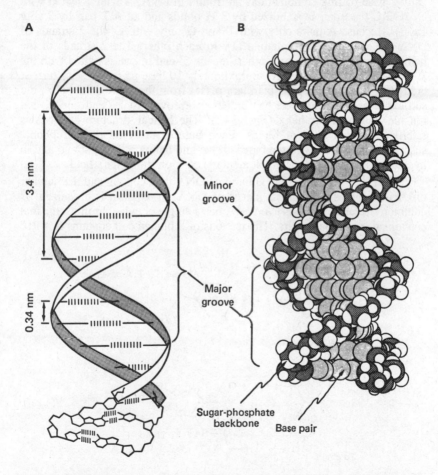

Figure 13–2. The Watson-Crick model of double-stranded DNA. *A:* Diagrammatic model. (Reproduced, with permission, from Kornberg A: *DNA Replication.* Freeman, 1980.) *B:* Space-filling model. (Redrawn, with permission, from Alberts B et al: *Molecular Biology of the Cell.* Garland, 1983.)

other end of the DNA (the 3' end), it is a 3' carbon that is nearest to the end of the polymer.

In most DNA molecules, the amount of dAMP (A) equals the amount of dTMP (T) and the amount of dGMP (G) equals the amount of dCMP (C). This observation, in combination with x-ray crystallographic data, prompted Watson and Crick to propose that DNA is **double-stranded.** In the Watson-Crick model of DNA, 2 continuous polymers of DNA are wrapped around each other in a double helix (Fig 13–2). The sugar and phosphate groups are on the outside of the molecule. In the center of the helix, the bases of the 2 DNA strands interact (pair) with each other through H bonds. Only 2 base-pairing combinations are found in DNA: A with T and G with C. A G:C base pair is stabilized by 3 H bonds and an A:T pair by 2 (Fig 13–3). Because A pairs only with T and G only with C, the 2 strands of DNA are said to be **complementary** to each other. The 2 strands of the helix are **antiparallel** to each other; ie, the 5' end of one is paired with the 3' end of the other. There are approximately 10 base pairs for every turn of the helix. Because the base pairs are offset from the axis of the helix, the surface of a DNA molecule has 2 distinct grooves, major and minor, that run the length of the helix (Fig 13–2). The helical structure of DNA is stabilized by noncovalent forces. Each base is involved in hydrophobic stacking interactions with the bases above and below it in the strand and in hydrogen bonds with the complementary base on the other strand.

Species differ both in the amount of DNA they contain and the number of **chromosomes** (continuous pieces of DNA) in which it is organized. A common bacterium, *Escherichia coli,* has a single circular chromosome that contains 4×10^6 base pairs. The nucleus of a human cell contains 6×10^9

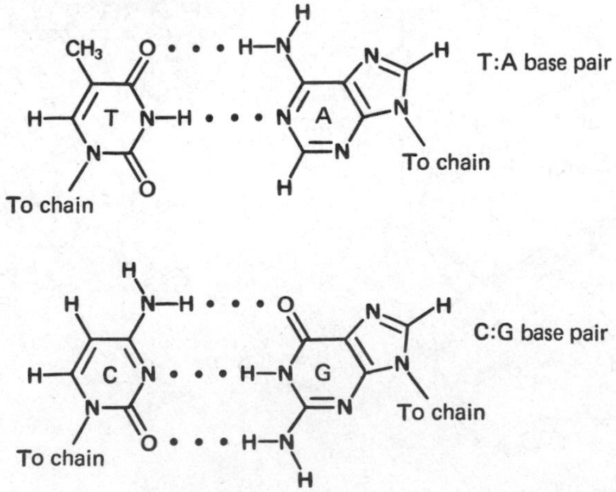

Figure 13–3. In DNA, A pairs with T and G pairs with C. Hydrogen bonds are represented by dotted lines. The base pairs are shown as if the observer were looking down the axis of the DNA helix.

base pairs of DNA, distributed among 23 pairs of linear chromosomes. Each human mitochondrion contains one small, circular chromosome. The nuclear chromosomes encode most of the proteins produced by the human cells. The mitochondrial chromosome contains instructions for the synthesis of some of the mitochondrial proteins. With few exceptions, every cell in a multicellular organism contains the same amount of DNA. The total DNA content of a cell is its **genome.** That portion of the DNA that codes for an individual function, eg, a polypeptide chain, is termed a **gene.**

Chromatin Structure

The DNA found in a single human cell has a total length in excess of 1 m, and yet a typical human cell is only 20 μm long. So that it can be packed into such a small space, the DNA must be condensed into a compact structure. Condensation is carried out by a number of proteins, and the condensed DNA-protein complex is called **chromatin.** In eukaryotes, the most abundant DNA-condensing proteins are the **histones,** a class of small proteins rich in arginine and lysine residues. These positively charged amino acids play an important role in binding to the negatively charged phosphate backbone of DNA. The structures of the histone proteins have been highly conserved throughout evolution. In fact, histones of humans differ only

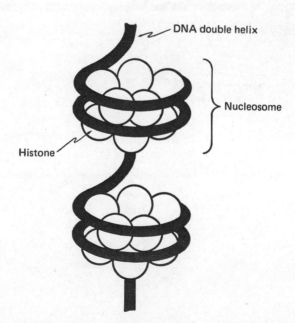

Figure 13–4. Model of nucleosome structure. (Redrawn and reproduced, with permission, from Löffler G et al: *Physiologische Chemie.* Springer-Verlag, 1979.)

slightly from those found in plants. Prokaryotes do not contain histones but do condense their DNA using histonelike proteins.

In eukaryotes, chromatin may be condensed as much as 8000-fold relative to naked DNA. The extent of condensation varies throughout the genome; those portions of the DNA not involved in synthesis of proteins are the most tightly condensed. The first step in condensation involves formation of **nucleosomes,** beadlike structures that contain 2 loops of DNA (146 base pairs) wrapped around a protein core (Fig 13–4). Each nucleosome core contains 8 proteins, 2 each of 4 histones—H2A, H2B, H3, and H4. Adjacent nucleosomes are separated by about 60 base pairs of DNA. A fifth histone, H1, ties the individual nucleosomes together in a closely packed string that can be further condensed to form a fiber 30 nm in diameter. Inactive DNA is condensed still further. In order to facilitate segregation of chromosomes to daughter cells, prior to mitosis the DNA is condensed to form the mitotic chromosomes visible by light microscopy.

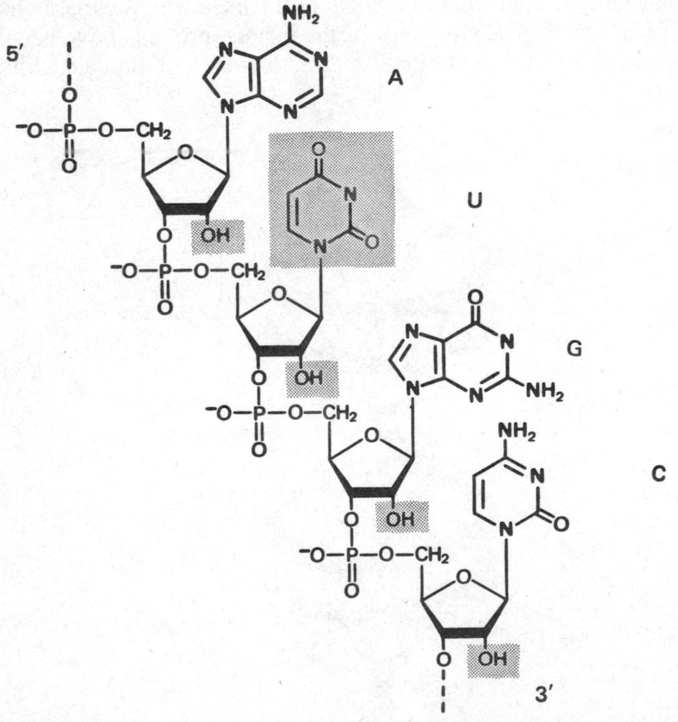

Figure 13–5. A segment of RNA. Shading indicates the features of RNA that distinguish it from DNA. The 4 bases of RNA are adenine (A), uracil (U), guanine (G), and cytosine (C).

RNA Structure

Although some viruses use double-stranded DNA as their genetic material, others use single-stranded DNA or a related nucleic acid, **RNA (ribonucleic acid)**, in either its single- or double-stranded form. RNA, like DNA, is a polymer of nucleoside monophosphates (Fig 13–5). It differs from DNA in that it contains ribose instead of deoxyribose and uracil instead of thymine. Uracil has the same base-pairing properties as T and can pair with A. Double-stranded RNA has a structure very similar to that of double-stranded DNA. By definition, a single-stranded nucleic acid is one that is not base-paired to a complementary strand throughout its length. However, all single-stranded nucleic acids do contain short helical regions that are formed by pairing between complementary segments of the same polymer (Fig 13–6).

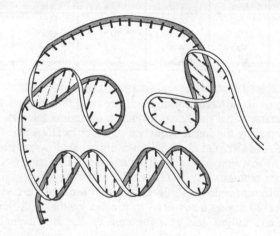

Figure 13–6. Single-stranded nucleic acids have short helical regions. (Reproduced, with permission, from Watson JD: *Molecular Biology of the Gene,* 3rd ed. Benjamin/Cummings, 1976.)

Genetic Code

Even before the structure of DNA was known, it was suspected that the purine and pyrimidine bases constituted the portion of the DNA molecule used to encode the amino acid sequences of proteins. This hypothesis was suggested by the observation that DNAs of all members of a given species have the same base composition, whereas DNAs of different species differ in base composition. It was also clear that there could not be a simple one-to-one code between bases and amino acids; DNA contains only 4 kinds of

Table 13–1. The genetic code (codon assignments in DNA).

First Position (5' End)	Second Position				Third Position (3' End)
	T	C	A	G	
T	Phe	Ser	Tyr	Cys	T
	Phe	Ser	Tyr	Cys	C
	Leu	Ser	Stop	Stop	A
	Leu	Ser	Stop	Trp	G
C	Leu	Pro	His	Arg	T
	Leu	Pro	His	Arg	C
	Leu	Pro	Gln	Arg	A
	Leu	Pro	Gln	Arg	G
A	Ile	Thr	Asn	Ser	T
	Ile	Thr	Asn	Ser	C
	Ile	Thr	Lys	Arg	A
	Met	Thr	Lys	Arg	G
G	Val	Ala	Asp	Gly	T
	Val	Ala	Asp	Gly	C
	Val	Ala	Glu	Gly	A
	Val	Ala	Glu	Gly	G

bases, whereas 20 different amino acids are used to construct proteins. We have since learned that the linear sequence of amino acids in a protein is, in fact, encoded by the linear sequence of bases in DNA and that the code found in DNA is a **triplet code;** ie, each amino acid is specified by a code word **(codon)** that consists of 3 adjacent bases. The triplet code of DNA uses a 4-letter alphabet (A, T, G, and C) and consequently has a total of 64 (ie, 4^3) different codons (Table 13–1). Sixty-one of the codons are used to encode the 20 amino acids used in protein synthesis. Three codons do not specify any amino acid and function as the stop signals in protein synthesis.

The genetic code is very nearly universal. The code shown in Table 13–1 is used by most prokaryotes and the nuclear chromosomes of most eukaryotes. The genomes of mitochondria depart from the general usage. In mammalian mitochondrial DNA, methionine is encoded by both ATG and ATA; tryptophan is encoded by TGA and TGG; and AGA and AGG, which in most genomes encode arginine, are used as stop signals.

Fig 13–7 demonstrates how the sequence of bases in one strand of DNA would encode a sequence of amino acids and illustrates several important features of the code. The triplets are read sequentially in the 5' to 3' direction, starting at a fixed point. The code is **unambiguous;** ie, each codon specifies only one amino acid. It is also **degenerate;** ie, more than one codon may code for the same amino acid. Our example indicates that codons TCT and AGT both encode serine. The code is also **commaless**—the codons are not

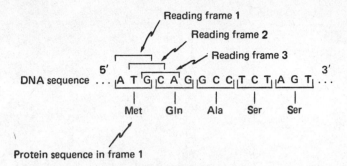

Figure 13–7. The genetic code is commaless, nonoverlapping, unambiguous, and degenerate.

separated from each other by noncoding nucleotides—and with rare exceptions, the code is **nonoverlapping**—each base in the DNA is part of only one codon. Because the code is commaless and nonoverlapping, in order to read the information, we must know which groups of 3 nucleotides constitute codons. In other words, we must select a **reading frame.** Because the genetic code is a triplet code, there are 3 possible reading frames on each strand. The mechanism by which the correct reading frame is selected is discussed below.

Degeneracy is one of the most important features of the code because it helps to minimize the effects of **mutations** (changes in the sequence of nucleotides in the DNA) (Fig 13–8). If the code were not degenerate, ie, if only 20 of the 64 codons were used to specify amino acids, most of the codons would signify nothing, which is interpreted by the protein synthesis machinery as a termination signal. If this were the case, most mutations in which one nucleotide was replaced by another (a substitution) would generate a noncoding codon (a **nonsense mutation**) and would lead to termination of the protein sequence. The code has evolved so that many substitution mutations have little or no effect on the structure of the encoded protein. Owing to the degeneracy of the code, many changes in the third position of a codon produce no change in the amino acid encoded (a **silent mutation**). Other single nucleotide changes lead to the replacement of one amino acid by another (a **missense mutation**) but often the new amino acid is similar to the one replaced—a conservative change. Insertions and deletions of nucleotides have more profound effects on the encoded protein than do substitutions. Because the code is commaless, if the number of nucleotides inserted or deleted is not a multiple of 3, the frame in which the code is read will be changed (a **frameshift mutation**) and most codons following the site of the mutation will be altered.

Both species and individuals depend for their survival on the ability to maintain and duplicate genetic material without changes in the encoded information. Mutations in the genome of germ line cells impair the ability

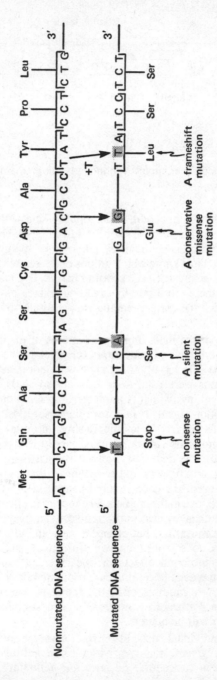

Figure 13–8. The form of the genetic code minimizes the effects of substitution mutations.

to produce viable offspring and thereby affect the survival of the species. Mutations in somatic cells appear to be involved in carcinogenesis. In fact, the information in genetic material changes relatively slowly. The human genome with 3.5×10^9 base pairs of DNA acquires on the average only 3 mutations per cell generation. The surprising stability of the genome depends upon a very accurate system of DNA synthesis (replication) and a number of highly active DNA repair systems.

DNA Replication

Before a cell can divide to form 2 daughter cells, it must first make a duplicate copy of its genetic material so that one copy can be distributed to each daughter cell. The process by which DNA is copied is **semiconservative;** ie, after replication, each of the daughter DNA molecules contains one old strand and one new strand. During replication, the duplex that is to be copied is locally separated into its component strands and each is used as the **template** for the synthesis of a new complementary strand (Fig 13–9).

Polymerization of new DNA chains is catalyzed by **DNA polymerase.** Bacteria possess 3 species of DNA polymerase: I, II, and III. Replication of the chromosome is catalyzed in large part by DNA polymerase III. The major role of DNA polymerase I is in DNA repair. The function of DNA polymerase II is currently unknown.

The substrates for DNA synthesis are deoxyribonucleoside 5'-triphosphates. As DNA polymerase moves along the template strand, substrate nucleotides pair with the template and are thereby selected for incorporation into the daughter strand. DNA polymerase can catalyze chain growth only in the 5' to 3' direction; ie, the incoming nucleotide is added to the 3' end of the growing chain (Fig 13–10). Formation of the phosphate ester linkages of the backbone involves the attack of the 3' hydroxyl of the last nucleotide of the growing chain on the α phosphate of the next substrate nucleotide. The β and γ phosphates of the substrate are released as pyrophosphate.

The fidelity of DNA polymerization depends on the accuracy with which the substrates form Watson-Crick pairs with the template. This accuracy is limited by the existence of rare tautomeric forms of the nucleotides that form base pairs other than the usual Watson-Crick pairs. Thymine, for example, can adopt an enol form that is unable to pair with adenine and instead pairs with guanine. Likewise, cytosine has an imino form that pairs with adenine (Fig 13–11). Although the lifetimes of these rare tautomeric forms are very short, they are long enough to permit incorrectly paired substrate molecules to be incorporated into DNA. Rare tautomers produce mistakes in polymerization in proportion to their abundance in the pool of nucleotides (about one part in 10^4). If these mistakes were uncorrected, the mutation rate would be unacceptably high.

The actual error rate in polymerization is much lower than one in every 10^4 nucleotides polymerized, because DNA polymerase **proofreads** its work

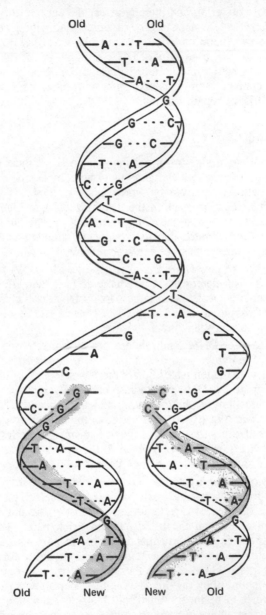

Figure 13–9. DNA is replicated semiconservatively. (Redrawn, with permission, from Watson JD: *Molecular Biology of the Gene,* 3rd ed. Benjamin/Cummings, 1976.)

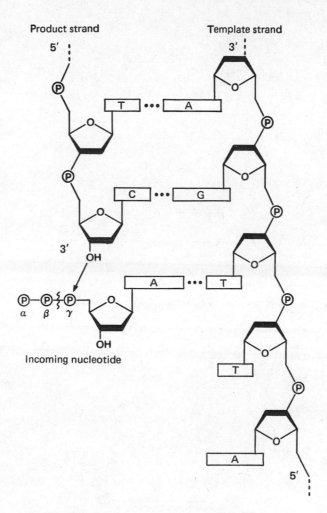

Figure 13–10. Synthesis of a new strand of DNA.

(Fig 13–12). Two properties of the enzyme are essential to its proofreading function: (1) DNA polymerase can add a nucleotide to the growing chain *only* if the preceding nucleotide is base-paired to the template, and (2) DNA polymerase has a 3′ exonuclease activity—ie, it can degrade a DNA polymer one nucleotide at a time starting from the 3′ end. The 3′ exonuclease is far more active on nucleotides that are not correctly paired with the template than on paired nucleotides. Because the lifetime of the rare tautomeric forms of the nucleotides is shorter than the time between polymerization steps, a misincorporated base will usually revert to its common tautomeric form

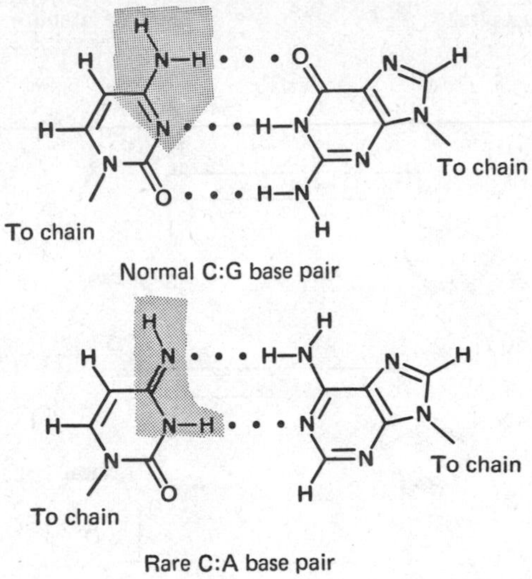

Normal C:G base pair

Rare C:A base pair

Figure 13–11. The imino form of cytosine pairs with adenine.

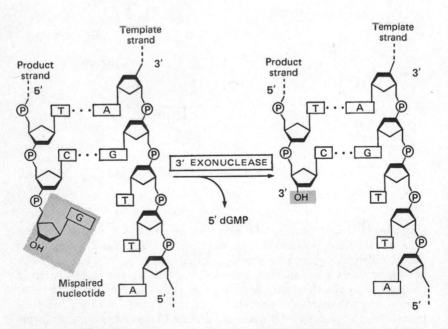

Figure 13–12. Proofreading involves removal of mispaired bases by a 3' exonuclease.

before the next base is added and in so doing will present an unpaired nucleotide to the exonuclease for removal. Removal of the incorrect base restores the 3' hydroxyl group of the preceding nucleotide. Another nucleotide can then be added to this end. With proofreading, DNA polymerase makes about one error for every 10^9 nucleotides polymerized.

Because DNA polymerase can only add a nucleotide to the 3' hydroxyl group of a nucleotide already stably paired with the template, the enzyme is unable to start new chains. How then is a chain initiated? The answer is that all new DNA chains begin with a short sequence of RNA, called a **primer** (Fig 13–13). The RNA primer is synthesized by a special RNA

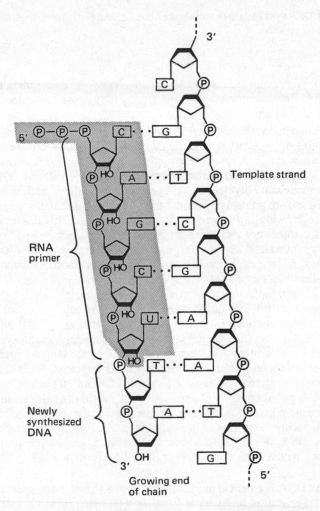

Figure 13–13. DNA synthesis is initiated by RNA primers.

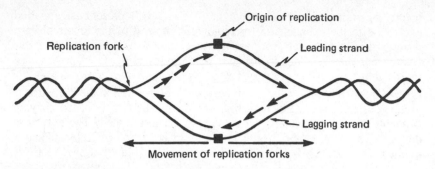

Figure 13–14. DNA synthesis is continuous on the leading strand and discontinuous on the lagging strand.

polymerase, primase, which is able to initiate chains and therefore is unable to proofread. Because primers contain errors, they must be replaced at a subsequent step in replication.

Replication of a chromosome is initiated at special signals in DNA, termed **origins of replication,** and the daughter chains grow bidirectionally from the origin on both template strands (Fig 13–14). Each origin thus generates 2 **replication forks** that move in opposite directions. Because DNA polymerase can polymerize chain growth only in the 5′ to 3′ direction, bidirectional growth poses a problem: 2 of the 4 growing chains at any origin appear to be growing in the wrong (3′ to 5′) direction. This dilemma was solved by the discovery that those strands that have a net 3′ to 5′ growth are actually synthesized as short fragments by DNA polymerases working 5′ to 3′. These fragments, termed **Okazaki fragments** after their discoverer, are later joined together to produce a continuous DNA molecule. Okazaki fragments of bacterial cells are approximately 1000 nucleotides long; those of eukaryotes are 100–200 nucleotides in length. At any replication fork, DNA synthesis is continuous on one strand (the **leading strand**) and discontinuous on the other (the **lagging strand**). On the leading strand, synthesis is initiated only once by an RNA primer, whereas on the lagging strand, an RNA primer is needed to start each new Okazaki fragment. Each Okazaki fragment begins with the synthesis of a primer of about 10 nucleotides (Fig 13–15). DNA polymerase then takes over and synthesizes the main body of the fragment. When DNA polymerase reaches the primer of the preceding Okazaki fragment, the primer is degraded by a 5′ exonuclease activity and replaced by DNA. Finally, the adjacent Okazaki fragments are joined by DNA ligase, which uses the energy of ATP to form a sugar-phosphate linkage.

Because DNA is a helical duplex in which one strand is wrapped around the other, its replication poses some complex topologic problems. For every 10 base pairs that are replicated, one turn of the parent helix must be

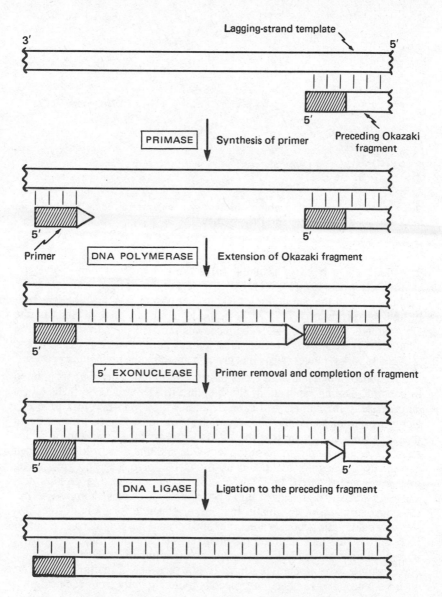

Figure 13–15. Steps in the synthesis of an Okazaki fragment.

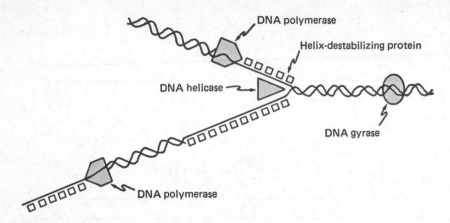

Figure 13–16. The topologic problems of DNA synthesis are solved by DNA helicase, helix-destabilizing protein, and DNA gyrase.

unwound. Three types of proteins solve the topologic problems of DNA synthesis (Fig 13–16). (1) DNA helicase acts at the replication fork to open the base pairs of the parent duplex. This energy-requiring reaction is driven by the hydrolysis of ATP. (2) A helix-destabilizing protein binds to the DNA strands unwound by DNA helicase and prevents them from re-forming a duplex. (3) DNA gyrase (also called topoisomerase) working ahead of the replication fork removes twists that are generated by the unwinding process.

Replication of eukaryotic DNA poses several problems not encountered in bacteria. During replication, the increase in DNA must be matched by an increase in the number of histones, and as the replication fork proceeds along the chromosome, the chromatin structure must first be unwound and then rewound. It appears that the histone core of the nucleosome remains attached to the leading strand of the template and that new histones are used to wrap up the lagging strand. Eukaryotic DNA polymerases progress along their templates about 10 times more slowly than do those of prokaryotes, and the Okazaki fragments of eukaryotes are 5–10 times shorter. The nucleosome structure of chromatin may be responsible for these differences. Because eukaryotic genomes are larger and because the polymerases progress more slowly, the time required for their replication would limit cell growth if replication were initiated from only one origin of replication on each chromosome. This problem is solved by the presence of multiple origins on each chromosome. Replication proceeds bidirectionally from each origin.

DNA Repair

Errors of replication are not the only threat to the stability of the genetic information. DNA is also subject to damage by a wide variety of chemicals and forms of energy. Thermal energy, for example, causes both deamination of adenine and cytosine residues and depurination. Ultraviolet light covalently links adjacent thymine residues. Ionizing radiation causes both single- and double-strand breaks. DNA-damaging events occur quite frequently and therefore could potentially result in a large number of mutations. Fortunately, most DNA damage is repaired. Repair depends upon 2 properties of DNA: (1) Its normal structure is distinct from the structures produced by DNA-damaging events, and thus the normal and abnormal can be distinguished. (2) DNA is double-stranded and thus contains a backup copy of the information it encodes. The importance of these 2 features of DNA for repair is illustrated in the following example.

Within the range of temperatures at which organisms can grow, cytosine residues spontaneously deaminate to form uracil. Because cytosine and uracil have different base-pairing properties, deamination changes the information encoded in the DNA. But because DNA does not normally contain uracil, the deamination product can be identified as a substrate for repair. Damage of this type is repaired by a process termed **excision repair** (Fig 13–17). First, a specific glycosylase recognizes and removes the damaged base. This

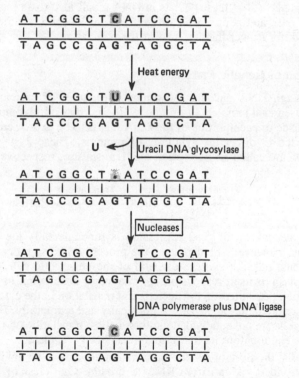

Figure 13–17. Uracil residues in DNA are removed by excision repair. (Courtesy of B Alberts.)

produces a site lacking a base. An endonuclease then makes a break in the damaged strand several bases to the 5' side of the site of damage. An exonuclease removes a short stretch of nucleotides; DNA polymerase fills in the gap by copying the other strand; and DNA ligase seals the remaining single-strand break. Excision repair is used to repair many kinds of damage that involve only one strand, eg, thymine dimers, missing bases, alkylations. In every case, the first step in repair is recognition of the abnormal structure by a specific repair enzyme.

Damage that involves both strands of a DNA duplex, such as chemical cross-linking or double-strand breaks, can also be repaired. The mechanisms involved in this type of repair are not well understood. In diploid organisms, recombination with the undamaged chromosome of a pair may be used to restore the original information of the DNA.

Several human diseases are known to result from defects in DNA repair. Individuals with **xeroderma pigmentosum** or **Bloom's syndrome** are defective in the repair of ultraviolet light damage. Those with **ataxia-telangiectasia** are abnormally sensitive to x-ray damage. Those with **Fanconi's anemia** are thought to be deficient in the repair of damage caused by cross-linking agents. All of these defects in DNA repair lead to an increase in the mutation rate, and people who suffer from any of these syndromes are at higher risk for cancer than the rest of the population.

Expression of Genetic Material

Although DNA encodes the primary structures of proteins, it is not directly used as a template in protein synthesis. Instead, a temporary RNA copy is made of each portion of the DNA that is to be expressed. That copy is then used to direct the synthesis of proteins. These 2 stages of gene expression are called **transcription** and **translation,** respectively.

$$\text{DNA} \xrightarrow{\text{Transcription}} \text{RNA} \xrightarrow{\text{Translation}} \text{Protein}$$

The mechanism of gene expression is fundamentally the same in all organisms. However, the details of the process differ significantly in prokaryotes and eukaryotes. The division of the eukaryotic cell into nucleus and cytoplasm is itself responsible for one major distinction. In eukaryotes, transcription takes place in the nucleus and translation in the cytoplasm. The 2 processes are therefore separated physically and temporally. Because prokaryotic cells are not subdivided into organelles, translation of a prokaryotic RNA can begin before its transcription has been completed.

In cells, the flow of genetic information appears to be unidirectional—from DNA to RNA and from RNA to protein. One class of viruses, the retroviruses, has a mechanism for reversing the first step in this flow. The retrovirus particle contains within it both its genetic material—a single-stranded RNA—and a viral enzyme, **reverse transcriptase.** The latter transcribes the genome to produce first a DNA-RNA hybrid double helix and then a DNA-DNA duplex, which can integrate by recombination into a host chromosome.

RNA Synthesis

At least 3 types of RNA are involved in protein synthesis. **Messenger RNA (mRNA)** acts as the informational copy of the gene that is to be expressed. **Transfer RNA (tRNA)** constitutes a class of adaptor molecules that bring the amino acids to the site of protein synthesis and read the mRNA codons. **Ribosomal RNA (rRNA)** is a part of the ribosome, which is a large assembly of protein and RNA that catalyzes the polymerization of the selected amino acids.

All 3 classes of RNA are transcribed from DNA by RNA polymerases. Not counting the primase that acts in replication, bacteria contain only one RNA polymerase. In eukaryotes, the genes for rRNA, mRNA, and tRNA are each transcribed by a different enzyme—RNA polymerase I, II, and III, respectively. Although they differ in the genes they transcribe, all of these enzymes catalyze the same type of reaction. Each uses ribonucleoside 5'-triphosphates as substrates and polymerizes RNA in the 5' to 3' direction. The reaction catalyzed by RNA polymerase is similar to that catalyzed by DNA polymerase; the polymer grows through the attack of the 3' hydroxyl group of the last nucleotide on the α phosphate of the incoming substrate nucleotide. Pyrophosphate is cleaved from all but the first nucleotide of the RNA chain. The template DNA directs the sequence of nucleotides in the RNA product by forming Watson-Crick base pairs with the substrate nucleotides. RNA synthesis therefore requires localized melting of the DNA duplex. The RNA product is complementary and antiparallel to its template. RNA polymerase has no proofreading function and thus is able to initiate chains.

RNA polymerase begins and ends transcription at discrete signals in the DNA called **promoters** and **termination signals,** respectively. These signals divide the DNA into **units of transcription.** Within any unit, only one strand of the DNA serves as template. In prokaryotes, promoter sequences are about 40 base pairs in length and are located immediately adjacent to the sequence that encodes the 5' end of the transcript (Fig 13– 18). Each promoter includes 2 highly conserved sequences located 35 and 10 nucleotides away from the transcriptional start. Because the latter is rich in A-T base pairs and is therefore stabilized by only 2 H bonds per base pair, it is thought to be the point at which the DNA duplex is opened by RNA polymerase.

In eukaryotes, each class of RNA polymerase recognizes a different type of promoter. RNA polymerase II promoters resemble those of prokaryotes. They are located adjacent to the sequences they control, and they contain several highly conserved sequences, including a sequence rich in

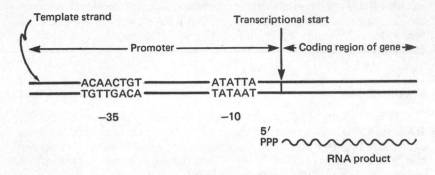

Figure 13–18. Bacterial promoters contain 2 highly conserved sequences located 35 and 10 nucleotides 5′ to the site of transcription initiation.

A-T base pairs. At least some promoters for RNA polymerase II can be read by *E coli* RNA polymerase. Surprisingly, the promoters of genes read by RNA polymerase III are found not adjacent to but within the transcribed sequence.

RNA polymerases are complex, multisubunit enzymes. That of *E coli* has 5 protein subunits, 2 α and one each of β, β', and σ (sigma). The complex of $\alpha_2\beta\beta'$ constitutes the catalytic core of enzyme. Sigma is a detachable subunit that functions only at the initiation of synthesis (Fig 13–19). To initiate transcription, the holoenzyme (core plus σ) binds to a promoter, melts open a short portion of the DNA duplex, and begins copying the template strand. Shortly after initiation, σ dissociates from the complex. While the core polymerase continues synthesis to the end of the transcription unit, the released σ subunit is recycled to another core polymerase. Termination signals are recognized by the core polymerase working either alone or in conjunction with a separate protein factor, ρ (rho).

RNA Processing

For many genes, the RNA produced by RNA polymerase, termed the **primary transcript,** must be altered via **RNA-processing reactions** before it can participate in protein synthesis. Processing may be used to make additions to the RNA chain (modification) or to remove blocks of nucleotides from it (nucleolytic processing). Processing plays a role in the synthesis of most eukaryotic mRNAs as well as tRNAs and rRNAs of both eukaryotes and prokaryotes. Prokaryotic mRNAs are synthesized without processing. In eukaryotes, RNA processing is performed in the nucleus.

Synthesis of a typical eukaryotic mRNA involves both modification and nucleolytic processing (Fig 13–20). Shortly after RNA polymerase II initiates synthesis of the RNA chain, the 5′-triphosphate end is modified by the addition of a **cap** structure. The cap consists of a 7-methylguanosine residue

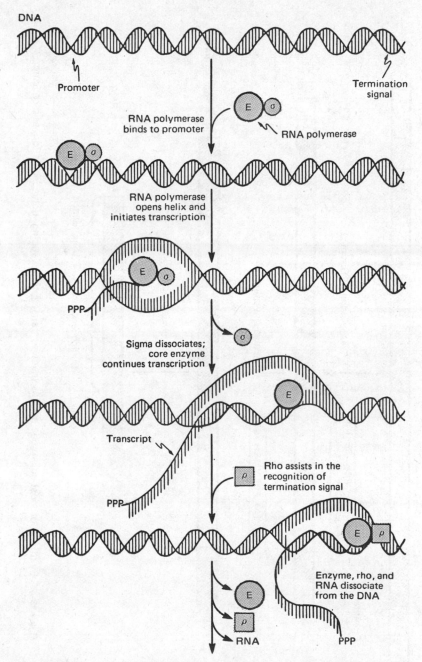

Figure 13–19. Sigma acts at the initiation of RNA synthesis and rho at termination. E = core enzyme.

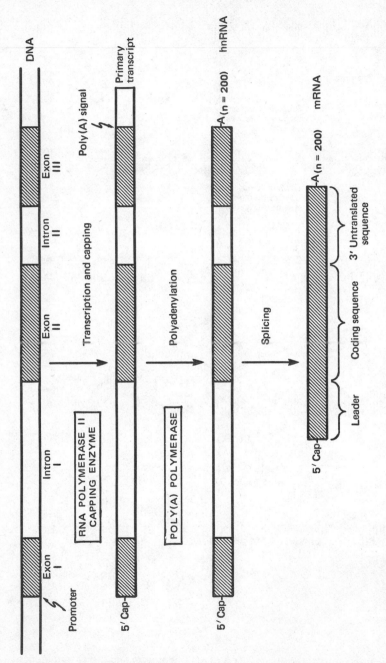

Figure 13–20. Synthesis of most eukaryotic mRNAs involves transcription, capping, polyadenylation, and splicing.

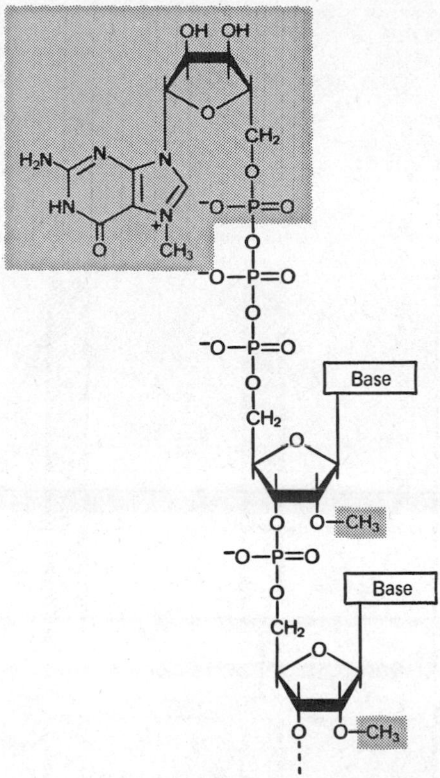

Figure 13–21. The 5′ cap structure of eukaryotic mRNAs.

joined to the first nucleotide of the RNA in an unusual 5′ to 5′ linkage (Fig 13–21). Some caps also include methyl groups attached to the 2′ hydroxyl group of the first 2 nucleotides of the chain. The cap of a eukaryotic mRNA plays an important role in the initiation of translation and may also protect the message from degradative enzymes and thereby increase its stability. The 3′ end of the primary transcript is also modified by processing enzymes. The original end is removed by an endonuclease and the new end modified by a template-independent poly(A) polymerase that adds a chain of 100–200 AMP residues. The function of **polyadenylation** is not yet clear. The capped and tailed transcript can be detected in the nucleus as one member of a class of molecules called **heterogeneous nuclear RNA (hnRNA).**

The function of the nucleolytic processing steps in the synthesis of eukaryotic mRNA is to assemble a continuous coding sequence. Most of the eukaryotic genes that encode proteins are substantially larger than the mRNAs they produce. The difference in size can be accounted for by blocks of noncoding nucleotides, termed **introns** or **intervening sequences,** that

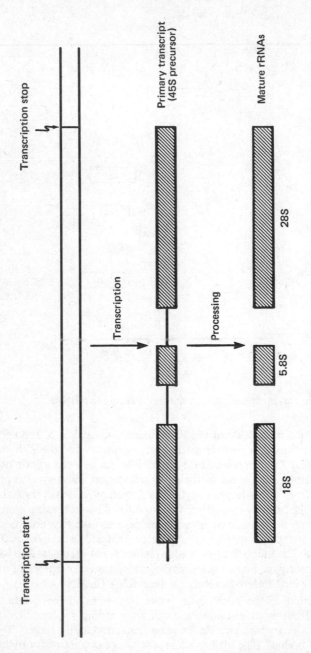

Figure 13–22. Three of the 4 rRNAs found in a eukaryotic ribosome are processed from a common precursor.

interrupt the coding sequences of these genes and divide them into coding blocks called **exons.** Transcription of an interrupted gene results in a primary transcript that contains both exons and introns. Removal of the introns, referred to as splicing, is the last step of processing. Splicing is catalyzed by enzymes that recognize signals located at the intron-exon borders, the splice junctions. These enzymes may be aided in their recognition of splice signals by an RNA-protein complex that contains a **small nuclear RNA (snRNA).** Some hnRNAs contain alternative sets of splice signals that permit splicing to form 2 different mature mRNAs that are translated to produce different but related proteins.

Following splicing, the mature mRNA is transported out of the nucleus to the site of protein synthesis, the cytoplasm. At this stage, the RNA has a 5' cap, a noncoding leader sequence that is not translated, a continuous coding sequence, a 3' untranslated sequence, and a poly(A) tail.

Transcripts of tRNA and rRNA genes of both prokaryotes and eukaryotes are also extensively processed. Some transcripts, such as those of rRNA genes and some prokaryotic tRNA genes, contain sequences corresponding to several mature RNAs (Fig 13–22). In those cases, processing is required to free the individual RNAs from the transcript. Even those tRNA transcripts that do not produce multiple mature species must be trimmed at both ends. Maturation of transcripts of tRNA and rRNA genes also requires modification of some nucleotides, forming minor nucleotide species. As many as 20% of the nucleotides of a tRNA may be modified by the addition of small groups (eg, methyl or isopentenyl). Modification of rRNA transcripts is less extensive.

Protein Synthesis

Protein synthesis is a complex process involving more than 100 different components. During protein synthesis, each messenger RNA is translated to form a polypeptide chain. This process requires the participation of a large number of proteins and RNAs, including tRNAs, aminoacyl-tRNA synthetases, ribosomes, and protein factors. Table 13–2 lists the molecules that participate in protein synthesis.

tRNA

Each mRNA contains a sequence of nucleotides that dictates the primary structure of a protein. However, mRNA is not able by itself to select the amino acids that are to be incorporated into the polypeptide chain. Because amino acids are too small to make contact with and "read" all 3 bases of a codon, another molecule must act as decoder of the message. This function is served by tRNA. Every cell contains more than 50 different species of tRNA, each of which can read one or more codons in mRNA.

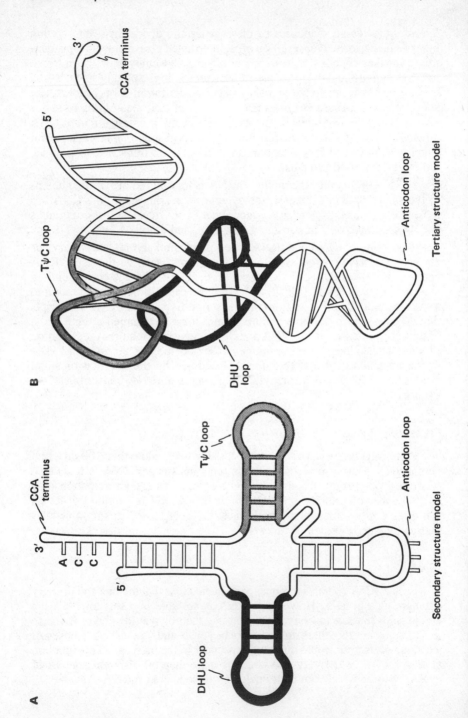

A

CCA terminus

3'
A
C
C

5'

DHU loop

TΨC loop

Anticodon loop

Secondary structure model

B

3'

CCA terminus

5'

TΨC loop

DHU loop

Anticodon loop

Tertiary structure model

Table 13–2. The roles of molecules involved in protein synthesis.

Molecule	Role
mRNA	Encodes protein primary structure
tRNA	Decodes mRNA and positions amino acid for peptide bond formation
Aminoacyl-tRNA synthetase	Charges tRNA with amino acid
ATP	Provides energy for tRNA charging
Ribosome	Organizes and catalyzes protein synthesis
Initiation factor	Promotes assembly of an initiation complex containing an mRNA, a ribosome, and an initiator tRNA
Elongation factor	Promotes translocation and binding of charged tRNAs to active ribosomes
Termination factor	In response to termination codons, dissociates the mRNA, the completed polypeptide, and the last tRNA from the ribosome
GTP	Provides energy for translocation and for function of initiation and elongation factors

Transfer RNAs are short polymers, 75–85 nucleotides in length. Each species of tRNA has a different primary structure (ie, sequence of nucleotides), but all fold into a common cloverleaf-shaped secondary structure, stabilized by Watson-Crick base pairs, and a common L-shaped tertiary structure, stabilized by additional contacts among several nucleotides (Fig 13–23). In addition to A, G, C, and U, tRNAs contain a large number of minor nucleotides located at characteristic sites in the molecule.

One end of the tertiary structure of tRNA consists of a short helix formed by pairing between the 5' and 3' ends of the polymer, with the 3' end extending 4 nucleotides beyond the end of the helix. All tRNAs contain the sequence CCA at their 3' ends. This sequence provides a site to which an amino acid can be covalently attached. Each species of tRNA joins specifically with only one species of amino acid. The other end of the tertiary structure consists of a loop formed by nucleotides located in the middle of the primary sequence. This loop contains a triplet of nucleotides, known as

---

Figure 13–23 *(at left).* The secondary *(A)* and tertiary *(B)* structures of tRNA. Shading is used to indicate the corresponding sections of the 2 models. The DHU loop contains 2–3 dihydrouridine residues and the TψC loop a thymine and a pseudouridine residue, all of which have arisen by posttranscriptional modification of U residues. (*B* is slightly modified and reproduced, with permission, from Stryer L: *Biochemistry,* 2nd ed. Freeman, 1981. Based on a drawing by Dr Sung-Hou Kim.)

the **anticodon,** that can base-pair with a codon in mRNA. During protein synthesis, each codon of the mRNA is read by a tRNA anticodon that can specifically base-pair with it. When it is base-paired to a codon, the tRNA positions its amino acid for incorporation into the growing polypeptide chain.

The anticodon of each tRNA is complementary to the codon it decodes and binds to it in an antiparallel fashion. Watson-Crick base-pairing rules govern the anticodon in pairing with the first (5') and second position of the codon. In reading the third codon position, however, the anticodon is able to **"wobble"** and can form both Watson-Crick and non−Watson-Crick base pairs. This portion of the tRNA is only loosely constrained by contacts with other nucleotides and is therefore relatively flexible. The base pairs that can be formed by anticodon wobble are listed in Table 13–3. As a result of wobble, a given tRNA can read more than one specific codon. For example, a single tRNA can wobble to read UCU and UCC, both of which code for serine. The organization of the genetic code ensures that wobble does not result in ambiguity of the genetic code. If, as a result of wobble, a tRNA can read multiple codons, all of those codons designate the same amino acid. As a consequence of wobble, organisms can economize on the number of tRNAs encoded and need not produce 61 different species of tRNA. However, in many cases more than one tRNA is needed to read all of the codons that signify a particular amino acid.

tRNA Charging

Each amino acid is attached to the 3' terminus of its tRNA via an ester linkage between the carboxyl group of the amino acid and either the 2' or the 3' hydroxyl group of the last nucleotide sugar. The job of correctly joining the tRNA to its specific amino acid, tRNA charging, is performed by a class of enzymes called aminoacyl-tRNA synthetases. There is one synthetase for each of the 20 amino acids. The reaction catalyzed by these

Table 13–3. Wobble pairs.

First Anticodon Base	Third Codon Base
C	G
A	U
U	A or G
G	U or C
Inosine*	U, C, or A

*Inosine, found in the wobble position of some tRNA anticodons, arises through the posttranscriptional modification of adenosine.

enzymes takes place in 2 steps (Fig 13–24). First, the amino acid is activated through the formation of an aminoacyl-AMP-enzyme complex. The enzyme then transfers the amino acid to the 3′ terminus of the appropriate tRNA. The bond formed between the amino acid and tRNA provides the energy required for the polymerization of the polypeptide.

Because tRNA serves as the interpreter of the mRNA, the accuracy of tRNA charging has a large effect on the overall accuracy of protein synthesis. In the charging step, accuracy is limited by the ability of the synthetase to discriminate between correct and incorrect amino acid–tRNA pairs. The inherent fidelity of the synthetases is quite low (about one mistake in every 100 tries). Their fidelity is, however, improved by proofreading. The synthetase scrutinizes the amino acid twice, once before activation and once after transfer to the tRNA. If the wrong amino acid has been attached to the tRNA, the aminoacyl-tRNA bond is usually hydrolyzed by the synthetase. Once the tRNA and the amino acid have been linked, it is the anticodon of the tRNA that determines where in the polypeptide chain the amino acid will be incorporated. A mischarged tRNA reads its usual codon and inserts the wrong amino acid into the sequence.

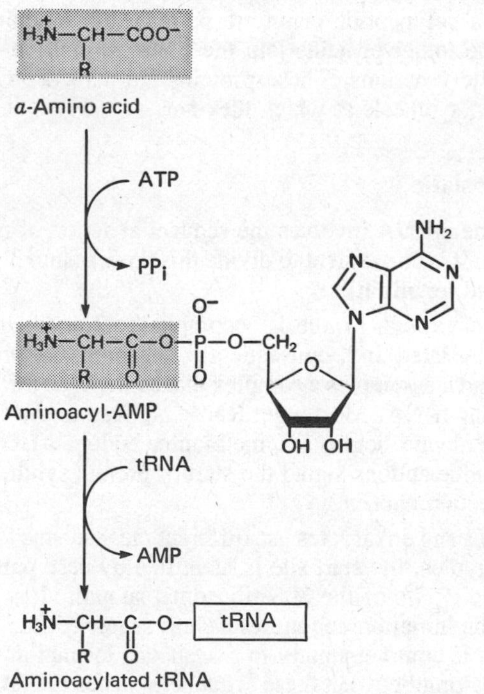

Figure 13–24. Amino acids are joined to the appropriate tRNAs in a 2-step reaction.

Ribosomes & Translation Factors

Protein synthesis takes place on the surface of a large assembly of proteins and RNA, the **ribosome**. Each ribosome contains a small number of specific RNAs (3 in prokaryotes and 4 in eukaryotes) and a large number of specific proteins (55 in prokaryotes and more in eukaryotes). Ribosomes both aid in the interaction between tRNA and mRNA and also provide the enzyme, peptidyl transferase, that catalyzes the peptide bond formation. Ribosomes do not play an informational role in protein synthesis.

Ribosomes of prokaryotes and eukaryotes differ in size and composition but are thought to be similar in architecture and mechanism of action. In both cases, the active ribosome is formed by the combination of 2 subunits of unequal size. Ribosomes and their subunits are named on the basis of the rate at which they move in a centrifugal field. Prokaryotic ribosomes are designated 70S (Svedberg units) and are made up of one 50S and one 30S subunit. Eukaryotes have 2 types of ribosome: cytoplasmic and mitochondrial. The mitochondria contain 70S ribosomes similar to those of bacteria. The cytoplasmic ribosomes are larger, ie, 80S with subunits of 60S and 40S. A number of drugs such as chloramphenicol, erythromycin, and tetracycline inhibit 70S but not 80S ribosomes.

While many of the proteins and RNAs needed to catalyze the polymerization of a polypeptide chain are permanently assembled in the form of the ribosome, other proteins join the ribosome only at discrete steps to mediate specific functions. These proteins are named on the basis of the stage in protein synthesis at which they act.

Events in Translation

Reading the mRNA involves the sequential action of many tRNAs and protein factors. It is convenient to divide this process into 3 steps: **initiation, elongation,** and **termination.**

Initiation of protein synthesis accomplishes 2 ends. It selects the first codon to be translated and, consequently, the frame in which the message is to be read, and it assembles a complex that contains an mRNA, a ribosome, and the initiating tRNA. Messenger RNAs are read in the 5' to 3' direction, and translation always begins at a methionine codon, AUG. However, only certain methionine codons signal the start of protein synthesis. How then is the initiating codon chosen?

Prokaryotes and eukaryotes use different mechanisms to accomplish this step. In prokaryotes, the start site is identified by base pairing between the mRNA and the rRNA of the 30S ribosomal subunit. Just 5' to each AUG that serves as an initiation codon lies a short sequence, the **Shine-Dalgarno sequence,** that is complementary to a sequence located at the 3' end of the rRNA. Base pairing between these sequences places the AUG in the correct site on the ribosome to be read as the first codon. A prokaryotic mRNA may contain more than one Shine-Dalgarno sequence and therefore may code for more than one protein. In eukaryotes, translation begins at the AUG closest to the 5' end of the mRNA. The ribosomes appear to bind to the end and then move along the chain to the first AUG codon. This mechanism allows selection of only one start site, and each eukaryotic mRNA therefore encodes only one protein.

There is only one codon for methionine, but each cell has 2 species of tRNA for decoding it. One species is used only in initiation and the other only in elongation of the polypeptide chain. In eukaryotes, the methionine tRNAs for initiation and elongation are designated Met-tRNA$_i$ and Met-tRNA$_m$, respectively. In bacteria, an N-formylated methionyl-tRNA, fMet-tRNA$_f$, is used to initiate synthesis. Formylation of the amino group takes place after charging of the tRNA (Fig 13-25).

Fig 13-26 illustrates the steps in the initiation of protein synthesis in prokaryotes. Aided by an initiation factor, the small ribosomal subunit binds to the mRNA. This step selects the initiation codon and sets the reading frame. Another initiation factor facilitates the binding of fMet-tRNA$_f$ and GTP to the complex. The final step in the formation of the initiation complex is the addition of the large ribosomal subunit, aided by a third initiation factor. This step results in the hydrolysis of the bound GTP and the release of the initiation factors. In the initiation complex, the fMet-tRNA$_f$ is paired with the AUG codon and is bound at a site on the ribosome designated the **peptidyl (P) site.** The next codon to be read is displayed at an adjacent site, the **aminoacyl (A) site,** which awaits the arrival of the next tRNA. Initiation of translation in eukaryotes differs from that in prokaryotes in at least 2 points. The small ribosomal subunit first binds the initiator tRNA and an initiation factor. It then binds to the 5' end of the mRNA and moves to the first AUG codon. ATP is hydrolyzed in this scanning process.

Fig 13-27 illustrates the steps of the elongation phase of protein synthesis. First, a tRNA corresponding to the second codon binds to the A site of the ribosome. Its binding is facilitated by an elongation factor, EF-Tu in prokaryotes and EF-1 in eukaryotes, which carries a molecule of GTP. After the factor has deposited its tRNA on the ribosome, GTP is hydrolyzed and the factor released. Peptide bond formation is catalyzed by peptidyl transferase, an integral part of the ribosome. This is accomplished by the transfer of the first amino acid from its tRNA to the free amino group of the second amino acid. Another elongation factor, EF-G in prokaryotes and EF-2 in eukaryotes, then triggers the release of the uncharged tRNA from the P site of the ribosome. Simultaneously, the second tRNA, bearing the dipeptide, is **translocated** from the A site to the P site and the ribosome moves to read the next triplet of the message. One GTP is split in the translocation step.

After 15-20 amino acids have been incorporated into the polypeptide chain, the N-formyl group and the amino terminal methionine may be removed. As soon as the ribosome has moved away from the initiation site, another may begin synthesis on the same mRNA. Translation of a single mRNA by several ribosomes leads to the formation of a **polyribosome.**

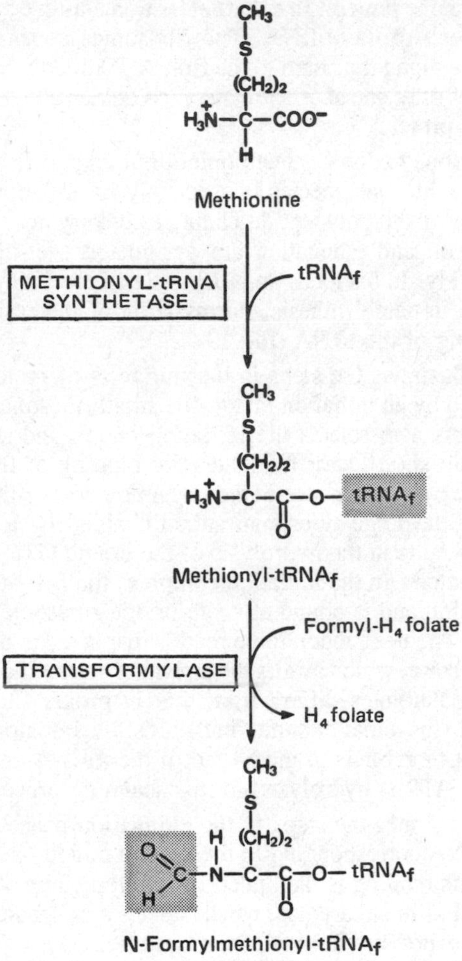

Figure 13–25. Formation of N-formylmethionyl-tRNA$_f$.

The 3 steps of elongation are repeated sequentially until a termination codon (UAA, UAG, or UGA) is displayed in the A site of the ribosome. In normal cells, there are no tRNAs that decode termination codons. The termination codon is read by a termination factor that stimulates peptidyl transferase to hydrolyze the bond between the last tRNA and the polypeptide chain. Release of the completed polypeptide chain is accompanied by the dissociation of the ribosomal subunits, the mRNA and the last tRNA. In prokaryotes, termination of protein synthesis does not require an energy source; in eukaryotes, GTP is hydrolyzed.

Regulation of Gene Expression

Every organism has the capacity to synthesize a large number of different proteins, and because those proteins are needed in different amounts and at different times, gene expression must be regulated. Gene expression may be regulated at the level of transcription of the DNA, at the level of translation or turnover of the mRNA, or, in eukaryotes, at the level of processing of the transcript or transport of the mRNA to the cytoplasm. Prokaryotic regulatory systems were the first to be studied and still provide the clearest understanding of regulatory mechanisms. However, recent years have seen major advances in the study of eukaryotic regulation. This work suggests that eukaryotes make use of some regulatory mechanisms similar to those found in prokaryotes as well as others unique to eukaryotes.

Prokaryotic gene expression is largely regulated at the level of transcription. In many cases, control of transcription is mediated by DNA-binding proteins. This type of regulation is illustrated by the following example—the *lac* operon of *E coli*.

Lac Operon

E coli are able to use a number of different sugars, such as glucose, lactose, galactose, and arabinose, as sources of carbon atoms needed for growth. The consumption of glucose is, however, more efficient than that of other sugars because its metabolism requires fewer enzymes. In order to economize on enzyme production, *E coli* synthesize those required for the catabolism of other sugars only when the sugar in question is present and glucose is absent. In order to metabolize lactose, *E coli* must synthesize 2 additional enzymes: β-galactosidase, which degrades the disaccharide lactose to its component sugars (glucose and galactose), and lactose permease, which transports lactose into the cell. When grown in the absence of lactose, bacteria produce only a few molecules of these enzymes per cell. But when the growth medium lacks glucose and contains lactose, production of the enzymes for lactose metabolism increases as much as 100-fold. The mechanism by which the synthesis of these proteins is controlled has been elegantly explained by the *lac* operon model proposed by Jacob and Monod. The features of this model are summarized below.

The genes that encode β-galactosidase *(z)* and lactose permease *(y)* are clustered together on the *E coli* chromosome with another gene *(a)*, that encodes an enzyme of unknown function (Fig 13–28A). Genes *z, y,* and *a* are transcribed from a single promoter and thus constitute a unit of transcription (an **operon**). Translation of the *lac* operon mRNA is initiated at 3 separate AUG codons that correspond to the first amino acid of each protein. Thus, the mRNA contains 3 units of translation (**cistrons**) and is referred to as a **polycistronic mRNA.** Expression of *z, y,* and *a* is controlled

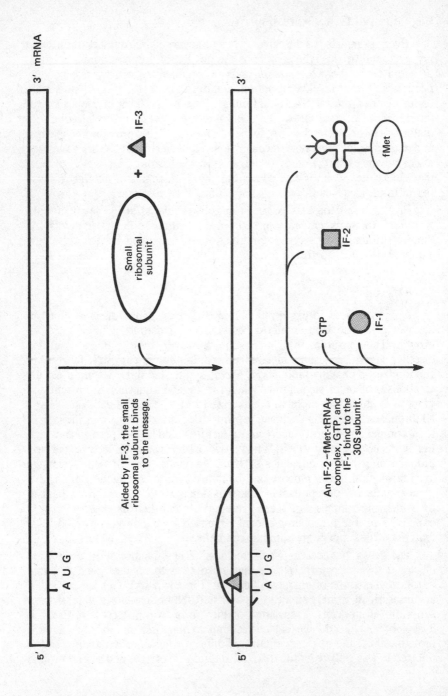

3' mRNA

AUG

5'

Aided by IF-3, the small ribosomal subunit binds to the message.

Small ribosomal subunit

+ IF-3

3'

AUG

5'

An IF-2–fMet·tRNA$_f$ complex, GTP, and IF-1 bind to the 30S subunit.

fMet

IF-2

GTP

IF-1

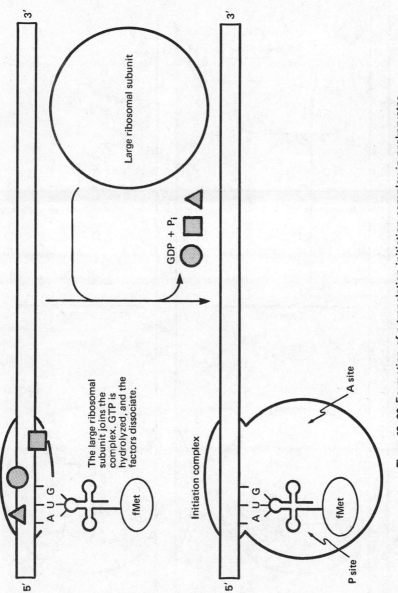

Figure 13–26. Formation of a translation-initiation complex in prokaryotes.

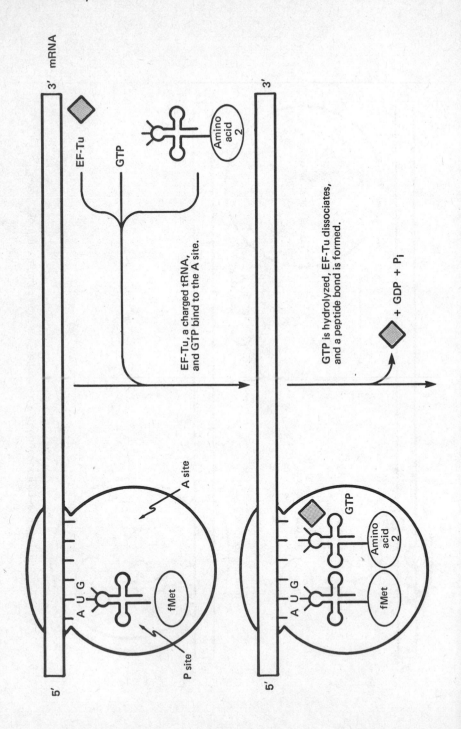

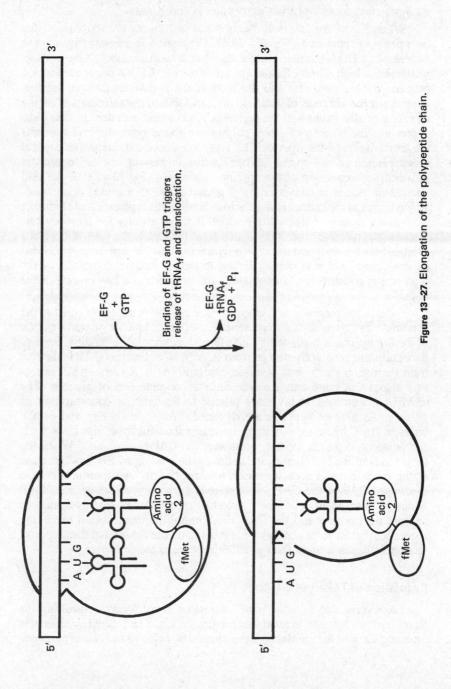

Figure 13–27. Elongation of the polypeptide chain.

at the level of initiation of transcription, and because a single mRNA encodes all 3 proteins, control of their expression is **coordinate**.

When *E coli* are grown in the absence of lactose, transcription of the *lac* operon is **repressed** (Fig 13–28B). Repression is mediated by the *lac* **repressor**, a DNA-binding protein that has 2 conformations. One conformation has a high affinity for the *lac* **operator** (o), a DNA sequence located between the *lac* promoter and the start of the β-galactosidase coding sequence. In the absence of lactose, the DNA-binding conformation of the repressor predominates and the repressor binds to the operator. In doing so, it prevents the binding of RNA polymerase to the promoter and prevents the transcription of the operon. The repressor thus exerts **negative control** on expression of the operon. When lactose is present, the *lac* operon is **induced**; ie, expression of the operon is increased (Fig 13–28C). A small amount of lactose is converted by β-galactosidase to a triose, allolactose, which serves as the **inducer**. Allolactose binds to the repressor and stabilizes it in its alternative conformation, which has low affinity for the operator. Because the repressor then dissociates from the DNA, the operon becomes available for transcription by RNA polymerase. The *lac* repressor is encoded by the *i* gene, which is closely linked to the *lac* operon. Synthesis of the repressor is **constitutive** (unregulated) and proceeds at a low rate all of the time.

Even in the presence of lactose, the *lac* operon is not transcribed at the maximal rate if glucose is also present. The effect of glucose on transcription of the *lac* operon is mediated by cAMP and another DNA-binding protein, the **catabolite gene activator protein (CAP)**. Note that this CAP is distinct from the one at the 5′ end of eukaryotic mRNA. In *E coli*, cAMP serves as a signal that represents the intracellular concentration of glucose. The cAMP concentration is inversely related to the internal concentration of glucose. As glucose supplies are depleted, cAMP levels rise and cAMP binds to the CAP. cAMP binding increases the affinity of the CAP for a site adjacent to the *lac* operon promoter, the CAP binding site. When the CAP binds to the *lac* operon, it facilitates the binding of RNA polymerase at the promoter and thus increases transcription of the operon. Because binding of CAP to the CAP site increases expression of the operon, CAP is said to exert **positive control**. Note that the regulatory signals of the *lac* operon, the operator and the CAP site, control the transcription of coding sequences that lie immediately adjacent to them. The function of these control sequences depends upon their position relative to the promoter.

Regulation of Eukaryotic Genes

Eukaryotes also regulate gene expression primarily by controlling the amount of mRNA that is available for translation. The following examples illustrate some of the regulatory devices used by eukaryotes for this purpose.

A. Structure of the *lac* operon

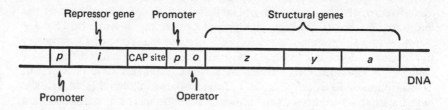

B. Repression: Lactose absent, glucose present

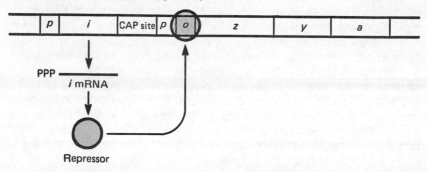

C. Induction: Lactose present, glucose absent

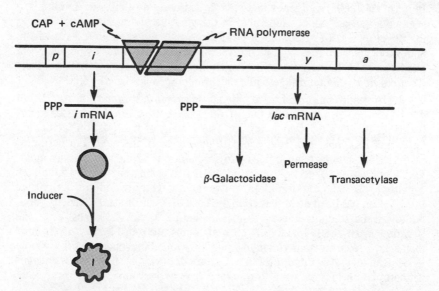

Figure 13–28. Structure and operation of the *lac* operon. The several elements of the operon are not drawn to scale.

It is likely that many eukaryotic genes are regulated by DNA-binding proteins which, like the *lac* repressor and the CAP, control the rate of transcription of those genes to which they bind. This form of control has been documented in the case of a primate virus, simian virus 40 (SV40). SV40 encodes a protein known as the T antigen that serves as a repressor for a class of SV40 genes called the early genes. Like the *lac* repressor, T antigen prevents mRNA synthesis by physically blocking the initiation of transcription by RNA polymerase. T antigen binds to the viral DNA at a site which partially overlaps with the promoter it controls. The T antigen binding site is therefore an operator sequence.

Eukaryotes also have a class of control sequences, called **enhancers,** at which positive regulatory proteins act to increase transcription. Enhancers differ from control sequences like the CAP binding site of *E coli* in that they need not be immediately adjacent to the coding sequences they control. An enhancer can increase the transcription from a promoter located as many as 1000 base pairs away. The mechanism by which these control elements act is as yet unknown. Enhancers apparently mediate the control of some genes under the control of steroid hormones. Steroid hormones affect their target cells via an intracellular hormone-binding protein, the hormone receptor. When the steroid hormone enters the cell, it binds to the receptor in the cytoplasm. The receptor-hormone complex then migrates to the nucleus, where it binds to the chromatin at specific sites, some of which are known to be enhancers.

Eukaryotic gene expression is also controlled by gene rearrangement. In germ line cells, the genes that encode antibody proteins are divided into segments that are widely separated in the genome and are inactive. During the development of an antibody-producing cell, the separate pieces of the antibody genes are relocated near each other and a promoter associated with one fragment is brought under the control of an enhancer adjacent to another. As a result of this rearrangement, the promoter is activated.

DNA modification may also play a role in transcriptional control. About 4% of the dCMP residues in mammalian DNA are methylated after incorporation, forming 5-methylcytosine. The pattern of methylation is not random, and it is inherited from one cell generation to the next. Furthermore, methylation patterns change during differentiation. Sequences that are methylated are less likely to be transcriptionally active than are those which lack 5-methylcytosine. A correlation between methylation and transcriptional inactivation has been documented in the case of X-chromosome inactivation. Although the somatic cells of women contain 2 X chromosomes, only one is active in any cell. Inactive X chromosomes are more highly methylated than active ones, and experimental demethylation of the inactive chromosome leads to an increase in its expression. These observations suggest that changes in methylation may be used to regulate gene expression.

Biologic Membranes | 14

OBJECTIVES

- Be able to describe the fluid mosaic model of a biologic membrane.

- Be able to describe the process by which membrane and secreted proteins are synthesized and delivered to the locations at which they function.

- Be able to describe how compounds are transported across membranes and what forms of energy are used to drive transport. Know how passive transport differs from active transport.

- Be able to describe the process of receptor-mediated endocytosis and to explain the role that coated pits, coated vesicles, and changes of pH play in this process.

BIOLOGIC MEMBRANES are nonpolar barriers that surround all living cells. Membranes regulate the flow of molecules and signals in and out of the cell and provide barriers across which ionic gradients can be established. In eukaryotes, membranes divide the internal space of the cell into several organelles—nucleus, cytoplasm, endoplasmic reticulum, Golgi apparatus, lysosomes, secretory vesicles, and mitochondria (Fig 14–1). This evolutionary development both increases the area that can be used for membrane functions and allows the cell to isolate one reaction from another. The membrane that surrounds a eukaryotic cell is termed its plasma membrane. Double membranes surround both the nucleus and the mitochondria. The outer membrane of the nucleus is continuous with that of the endoplasmic reticulum.

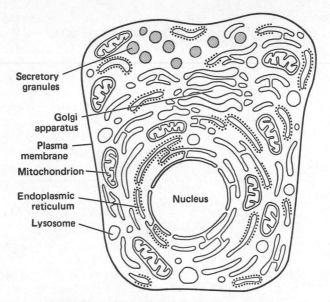

Figure 14–1. The structure of a eukaryotic cell.

Membrane Structure

All biologic membranes contain both lipids and proteins. The major function of the lipids is the formation of a nonpolar barrier. Membrane proteins catalyze reactions occurring at the membrane and mediate selective communication across it. Fig 14–2 illustrates the current model of membrane structure—the **fluid mosaic model.**

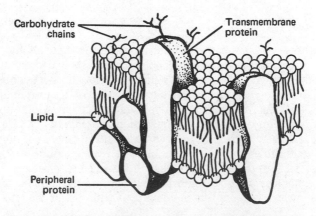

Figure 14–2. The fluid mosaic model of membrane structure.

The basis of all biologic membranes is a bilayer of amphipathic lipids arranged with the polar portion of each molecule at one of the 2 surfaces of the bilayer and the nonpolar portion in the core. Membranes contain many different lipids, including phospholipids, cholesterol, sphingolipids, and glycolipids (see Chapter 9). Cells differ in their lipid compositions, and even the 2 faces of a membrane differ. For example, the plasma membrane of a red blood cell contains predominantly phosphatidylserine and phosphatidylethanolamine on the cytoplasmic surface and sphingomyelin and phosphatidylcholine on the noncytoplasmic surface. In all cells, only the noncytoplasmic surface of the membranes contains glycolipids. Thus, membranes are asymmetric. It is not yet clear why more than one type of lipid is needed or what determines the proportions of the various lipids in a given membrane.

Assembly of the lipids into a bilayer is favored by hydrophobic interactions among the nonpolar portions of the lipid molecules. The lipids are free to diffuse laterally within the plane of the membrane, but because the nonpolar phase of the membrane is a substantial barrier to polar groups, lipids only slowly flip from one side of the membrane to the other. Most of the phospholipids of the membrane contain one saturated and one unsaturated fatty acid chain. Because the unsaturated fatty acids are bent, they disrupt the regular packing of the lipids within the bilayer and increase its fluidity.

Proteins that are embedded within the lipid bilayer are referred to as **integral membrane proteins.** Those which extend through the entire thickness of the membrane are designated **transmembrane proteins. Peripheral membrane proteins** are those attached to one surface of the membrane by contacts with other proteins. Nonpolar amino acids predominate in the portion of a membrane protein that extends into the lipid bilayer. In many cases, the portion of a transmembrane protein that crosses the bilayer has a secondary structure consisting of an α helix or a β-pleated sheet. These structures have the advantage of providing hydrogen-bonding partners for all of the hydrogen-bonding groups of the polypeptide backbone as it traverses the bilayer. Many membrane proteins contain chains of sugars (oligosaccharides) attached covalently to those portions of the polypeptide that are exposed on the noncytoplasmic face of the membrane.

Some membrane proteins are free to move about within the plane of the membrane, while others are fixed in one position. The latter are attached to networks of proteins either inside or outside the cell. The anion channel of red blood cells, called the band 3 protein, is an example of a protein that can be anchored. Some of the band 3 molecules are bound to a network of proteins that makes up a submembrane cytoskeleton of the cell (Fig 14–3). Many cells that are embedded within a solid tissue are attached to the extracellular matrix or to neighboring cells through contacts with membrane proteins.

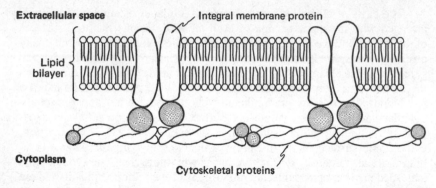

Figure 14–3. The red blood cell anion channel is attached to the cytoskeleton.

Membrane Biosynthesis

As the cell grows, the membrane surface is expanded by the addition of new components to the existing membrane of the endoplasmic reticulum. The endoplasmic reticulum has 2 histologically distinct components—smooth and rough. Most of the enzymes that catalyze the synthesis of membrane lipids are located on the cytoplasmic surface of the smooth endoplasmic reticulum, and newly synthesized lipids are inserted into the bilayer at that site. Because new lipids appear on both faces of the endoplasmic reticulum membrane, it is probable that this organelle has a mechanism for flipping lipids from the cytoplasmic face to the luminal face. The enzymes that add the sugars to glycolipids are found in the lumen of the Golgi apparatus. Thus, glycolipids need never appear on the cytoplasmic surface of the membrane.

Secreted proteins, proteins found in the interior of the lysosomes, and many integral membrane proteins are synthesized at the rough endoplasmic reticulum. The ribosomes engaged in the synthesis of these proteins are bound to the cytoplasmic surface of the membrane and account for its rough appearance. Translocation of the newly formed protein across the membrane takes place as it is being polymerized; ie, translocation is **cotranslational** (Fig 14–4). Attachment of ribosomes to the membrane and insertion of the protein depend on the presence of a **signal sequence** that is part of the growing polypeptide chain. The process begins with the assembly of a translation initiation complex using ribosomes that are free in the cytoplasm. Translation of the 5' portion of the message results in the synthesis of the N-terminal portion of the protein, which contains a signal sequence—a sequence of hydrophobic amino acids generally 20–25 amino acids long. As soon as the signal is produced, it is bound by a complex of proteins and RNA—the **signal recognition particle (SRP)**—and translation is temporarily arrested. The ribosome and the SRP then bind to a receptor protein

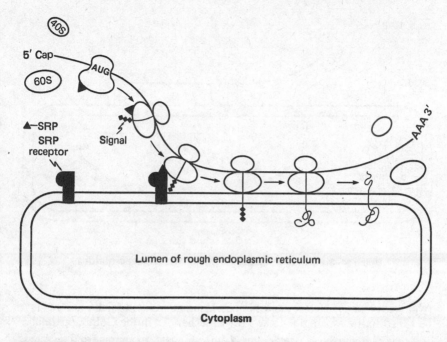

Figure 14–4. Cotranslational insertion of a protein into the membrane of the rough endoplasmic reticulum. (Redrawn, with permission, from Walter P, Gilmore R, Blobel G: Protein translocation across the endoplasmic reticulum. *Cell* 1984;**38**:5.)

that is located on the surface of the rough endoplasmic reticulum membrane, and translation resumes. As the protein is elongated, it passes through the membrane into the lumen. Lysosomal and secretory proteins pass completely through the membrane; in the case of membrane proteins, translocation stops while the protein still spans the lipid bilayer. Many of the proteins synthesized at the rough endoplasmic reticulum are proteolytically processed when they reach the lumen. There, the signal sequence may be removed by a specific protease, **signal peptidase.**

The initial steps in protein glycosylation accompany translocation (Fig 14–5). First, a treelike polymer of sugars is constructed within the lumen of the rough endoplasmic reticulum. **Dolichol,** a membrane lipid, serves to anchor the tree to the membrane during its construction. The tree, which is composed of mannose, glucose, and N-acetylglucosamine, is built up stepwise from nucleotide-activated sugars such as GDP-mannose. As the newly synthesized protein passes into the lumen of the rough endoplasmic reticulum, the oligosaccharide tree is transferred as a unit from the dolichol anchor to the amino group of an asparagine side chain. Following transfer of the carbohydrate to the protein, the glucose residues and some of the mannose residues are trimmed away. Additional sugars, including N-ace-

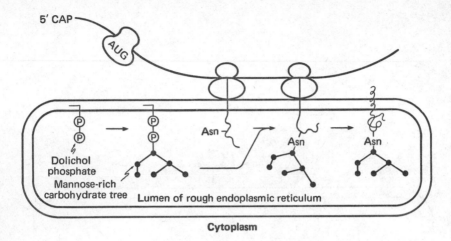

Figure 14–5. Formation of asparagine-linked oligosaccharides.

tylglucosamine, galactose, and N-acetylneuraminic acid (sialic acid), may be added to the asparagine-linked oligosaccharides in the Golgi apparatus.

In addition to the asparagine- or N-linked sugars, membrane and secreted proteins also contain glucose and galactose units attached to serine and threonine residues. Because they are linked through the hydroxyl groups of the amino acid side chains, the latter are termed O-linked sugars. These modifications are made in the Golgi apparatus.

What is the role of the oligosaccharides found on membrane and secreted proteins? It is clear that at least some proteins require glycosylation to stabilize their tertiary structures. If the addition of sugars is prevented, some of the newly synthesized proteins fold improperly and precipitate in the lumen of the rough endoplasmic reticulum.

The preceding section accounts only for growth of the membrane of the endoplasmic reticulum. Many other membranes within the cell grow by acquiring pieces of the endoplasmic reticulum membrane, a process that involves vesicle formation and fusion (Fig 14–6). The endoplasmic reticulum membrane regularly buds off vesicles that migrate to and fuse with the membranes of the Golgi apparatus. The Golgi apparatus uses the same process to distribute membrane materials to the lysosomes, secretory granules, and plasma membrane. Through exchanges of this kind, the interior surface of an organelle membrane can become the exterior surface of the plasma membrane.

Synthesis of mitochondrial membrane proteins is different. While some mitochondrial proteins are encoded by genes located on the mitochondrial chromosome and synthesized within the mitochondria, most are encoded by nuclear genes. Messenger RNAs transcribed from the latter are translated

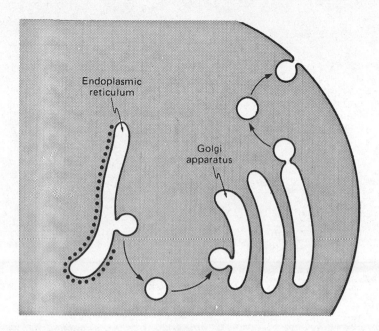

Figure 14–6. New membranes are distributed throughout the cell by a process of vesicle formation and fusion.

by free cytoplasmic ribosomes. Synthesis of the protein is completed before transport of the protein into the mitochondrion is initiated; ie, transport is **posttranslational.** Entry of proteins into the mitochondrion is mediated by a receptor protein located on the surface of the outer mitochondrial membrane and a signal sequence distinct from that used to direct proteins to the rough endoplasmic reticulum. The transmembrane proton gradient of the mitochondrion provides the energy needed for transport.

Secretion & Endocytosis

Cells take up proteins and particles from the environment through a process called **endocytosis** and secrete proteins using a related process called **exocytosis.** One form of endocytosis, receptor-mediated endocytosis, is illustrated in Fig 14–7. Embedded within the plasma membrane of all cells are receptor proteins, such as the LDL receptor (see Chapter 9), to which proteins and particles that are to undergo endocytosis specifically bind. These receptors are clustered within specialized regions of the plasma membrane termed **coated pits.** The cytoplasmic surface of the plasma membrane in a coated pit is covered by a geodesic dome–like structure made up of a protein named **clathrin.** Periodically, a coated pit invaginates to form a coated vesicle also surrounded by a clathrin coat. The vesicle then loses its clathrin

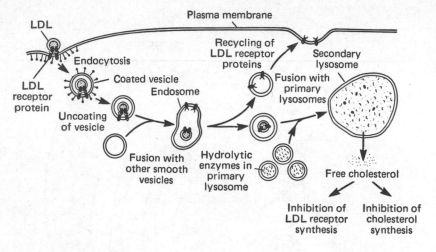

Figure 14-7. Receptor-mediated endocytosis of the LDL particle. (Reproduced, with permission, from Alberts B et al: *Molecular Biology of the Cell.* Garland, 1983.)

coat and fuses with another type of vesicle to form an endosome. Proton pumps located in the membrane of the endosome acidify its interior. In this acid environment, LDL dissociate from their receptors, which are recycled to the plasma membrane, and proceed to a lysosome, where they are catabolized.

A similar process, exocytosis, is used to deliver proteins made in the endoplasmic reticulum and Golgi apparatus to the outside of the cell. Some proteins are secreted continuously by a process termed **constitutive exocytosis.** Others are accumulated within the cell in specialized secretory granules and released to the outside only when their secretion is triggered by a specific signal. In the constitutive pathway, membrane vesicles carry the secretory proteins directly from the Golgi apparatus to the plasma membrane, where they fuse and release their contents. Proteins that are to be stored for later secretion are first delivered to condensing vacuoles and then to secretory granules.

Transport Across Membranes

Although water, gases, and some small compounds can cross membranes unaided, most substances, especially polar compounds, cannot. The latter are transported in and out of the cell by specific proteins. There are 2 forms of membrane transport: passive and active. The properties of transport systems are summarized in Table 14-1.

Passive transport is mediated by transmembrane proteins that facilitate the diffusion of substances across the lipid bilayer by providing specific

Table 14—1. Membrane transport systems.

Type of Transport	Source of Energy	Type of Protein
Passive	Electrochemical gradient of substance to be transported	Carrier or channel
Active	ATP or electrochemical gradients of Na^+ or H^+	Pump

carriers or channels. Passive transport is driven by the gradient of the substance that is being transported; ie, the substance moves from an area of higher to lower concentration. Consequently, no additional source of energy is required. The 2 types of proteins involved in passive transport, carriers and channels, differ in their selectivity and speed. **Carriers** are proteins that have specific binding sites for the substances they transport. They are highly selective and carry a small number of molecules each time they operate. As the concentration of the transported substance increases, the carrier may become saturated. **Channels** are relatively less selective and may transport several species that are of similar size and charge. Some channels, like the band 3 protein of red blood cells that transports chloride and bicarbonate ions, are open continuously. Others, such as calcium ion channels of muscle cells, open only in response to a specific signal (in that case, depolarization of the membrane) and are therefore called **gated channels.**

Active transport is performed by membrane proteins that act as pumps. Membrane pumps are powered by sources of energy other than the concentration gradient of the species they transport and are therefore able to transport substances from an area of lower to higher concentration. All pumps use one of 2 forms of energy: ATP or an ion gradient. Three types of pump are powered directly by ATP: H^+ pumps, Ca^{2+} pumps, and an Na^+/K^+ pump. Other pumps are driven by the Na^+ and H^+ gradients that are established by the ATP-driven pumps.

The Na^+/K^+ pump, known as Na^+/K^+-ATPase, is the best-studied membrane pump. This pump, located in the plasma membrane, pumps Na^+ out of and K^+ into the cell. The pump consists of a complex of 2 copies each of 2 polypeptides. One is a glycoprotein of unknown function. The other is a transmembrane protein that contains the sites at which Na^+, K^+, and ATP bind. Na^+ and ATP bind on the cytoplasmic surface and K^+ on the extracellular surface. The pump protein can be phosphorylated by ATP when Na^+ is bound. Although the mechanism of action of this pump is not yet understood, it is thought that pumping results from a conformational change that is induced by phosphorylation. Fig 14–8 illustrates how this might work. When Na^+ binds to a site on the cytoplasmic side of the pump protein, the protein can be phosphorylated by ATP. Phosphorylation produces a conformational change that transports Na^+ to the extracellular surface of the membrane and creates a K^+ site on the extracellular surface. Binding of K^+

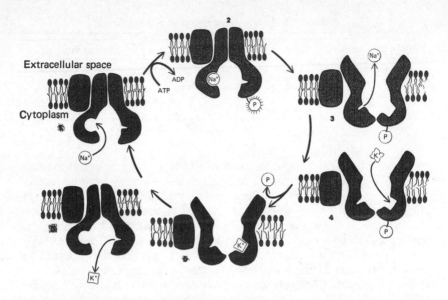

Figure 14–8. Na$^+$ and K$^+$ are transported across the plasma membrane by an ATP-driven pump. (Reproduced, with permission, from Alberts B et al: *Molecular Biology of the Cell.* Garland, 1983.)

then activates a phosphatase activity that removes the phosphate group from the pump protein. The protein reverts to its original conformation, bringing the K$^+$ site to the cytoplasmic side of the membrane. For every ATP hydrolyzed, 3 Na$^+$ and 2 K$^+$ are pumped across the membrane.

Na$^+$/K$^+$-ATPase can be inhibited by a class of drugs that increase the force of cardiac contraction, the cardiac glycosides. Drugs of this class, which includes **digitalis** and **ouabain,** compete with K$^+$ for binding on the extracellular surface of the protein and inhibit the action of the pump by preventing its dephosphorylation.

Membrane pumps that are driven by electrochemical gradients are used to transport a wide variety of substances, including ions, amino acids, and sugars. In every case, these pumps are driven by the major electrochemical gradient that spans the membrane in question. Bacteria, mitochondria, and intestinal cells have a substantial transmembrane H$^+$ gradient. Transport across the plasma membrane of many eukaryotic cells is driven by the Na$^+$ gradient that is established by the Na$^+$/K$^+$-ATPase.

Pumps powered by ionic gradients are classified as **symports** or **antiports,** depending on whether the 2 substances being transported are traveling in the same or opposite directions across the membrane (Fig 14–9). Glucose is transported from the lumen of the intestine into intestinal cells by a H$^+$-driven symport protein. Ca^{2+} is carried out of cardiac muscle cell by several

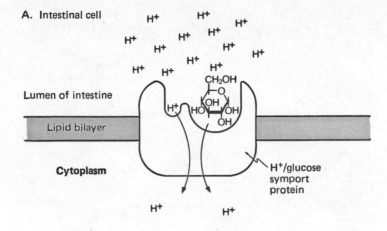

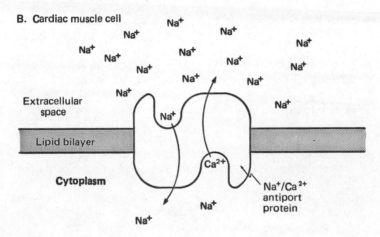

Figure 14–9. Pumps driven by ionic gradients are classified as either symports or antiports.

mechanisms, one of which involves a Na^+/Ca^{2+} antiport. The latter is driven by the gradient of Na^+ that is produced by the action of Na^+/K^+-ATPase. The existence of a cardiac Na^+/Ca^{2+} antiport suggests a mechanism by which the cardiac glycosides could increase the force of cardiac contraction. In the heart, the force of contraction increases with the concentration of Ca^{2+} in the cytoplasm (see Chapter 15). Because cardiac glycosides inhibit Na^+/K^+-ATPase, they decrease the Na^+ gradient across the membrane. This in turn decreases the rate at which the Na^+/Ca^{2+} antiport removes Ca^{2+} from the cardiac cell.

15 | Biologic Movement

OBJECTIVES

- Be able to describe the structure of the contractile unit of muscle tissue and to explain the mechanism by which it contracts.

- Be able to describe the role of ATP in contraction and to explain how ATP supplies are maintained in contracting muscle.

- Be able to explain how contraction is regulated in each of the 3 muscle types.

- Be able to describe the effects of epinephrine on cardiac and smooth muscle and to explain how those effects are mediated.

ALL CELLS are capable of movement. Muscle cells contract and, in doing so, move other tissues to which they are attached. Fibroblasts and other cells migrate with ameboid movements. Sperm cells propel themselves by the movement of flagella. In every case, movement involves the conversion of chemical energy (ATP) to mechanical work. The aim of this chapter is to describe how this transformation occurs. Because muscle is the best understood of all the biologic systems for movement, it will serve as the focal point for this discussion.

Structure of the Contractile Apparatus

Vertebrates contain 3 types of muscle: skeletal, cardiac, and smooth muscle. While each type is distinct in appearance and properties, all make use of the same mechanism of contraction. In order to explain that mechanism, it is necessary to describe the structure of muscle tissue. Of the 3 types of muscle, skeletal muscle is the most regular in its organization and thus the most easily studied and described.

A typical skeletal muscle is made up of a number of individual muscle fibers (Fig 15–1). Each fiber is a large, multinucleated cell, 20–100 μm

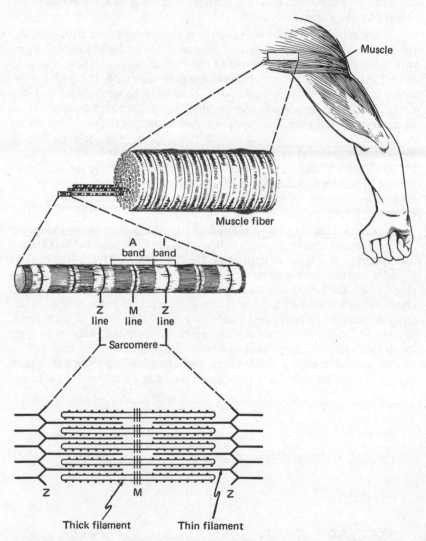

Figure 15–1. The structure of skeletal muscle. (Redrawn after Sylvia Colard Keene. Modified and reproduced, with permission, from Bloom W, Fawcett DW: *A Textbook of Histology,* 10th ed. Saunders, 1975.)

in diameter. The cytoplasm of the muscle cell (the **sarcoplasm**) contains a number of fibrils (**myofibrils**) that are organized in parallel arrays and run from one end of the cell to the other. The myofibrils exhibit a regular pattern of alternating dark and light bands, termed the A and I bands, respectively. Each I band is bisected by a Z line and each A band by an M line. This pattern of bands and lines is produced by a repeating array of **sarcomeres,** the basic contractile units of the muscle. The ends of the sarcomere are defined by the Z lines.

The sarcomere is made up of 2 types of interdigitating filaments: the **thick and thin filaments.** The thin filaments are anchored to the Z lines, pass through the I bands, and end in the central A band. Thick filaments extend from one end of the A band to the other and are held together at its center by the M line. In the outer portions of the A band, the thick and thin filaments overlap, and small cross-bridges project outward from the thick filaments toward the thin filaments. A cross section of the myofibril shows that the thick filaments are packed in a triangular array in which each thick filament is surrounded by 6 thin filaments. In skeletal and cardiac muscle, the sarcomeres of adjacent myofibrils are in register and give the muscle its striated appearance. Smooth muscle contains thick and thin filaments, but because they are not organized in regular sarcomeres, smooth muscle is not striated.

The thick filaments are composed of **myosin,** a large protein made up of 6 polypeptide chains: 2 identical heavy chains (MW about 200,000) and 2 copies each of 2 types of light chain (MW 15,000–20,000). The 6 polypeptides together form a structure that has a rigid tail and 2 globular heads (Fig 15–2). The tail consists of the carboxy-terminal portions of the 2 heavy chains wound around each other. Each globular head contains the amino-terminal portion of a heavy chain and one each of the 2 types of light chain. Myosin molecules spontaneously aggregate via their tail portions to form the thick filaments. The myosin molecules at either end of the filament are in opposite orientation to each other, and therefore the thick filament is bipolar. The globular heads of the myosin molecule correspond to the cross-

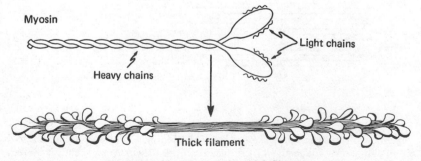

Figure 15–2. Myosin and the thick filament.

bridges that project outward from the thick filaments. While all myosin molecules have the same general architecture, distinct types of myosin are found in skeletal, cardiac, and smooth muscle. The differences in myosin molecules partially account for the different properties of the 3 types of muscle.

The major component of the thin filament is **actin**, a globular protein that contains a single polypeptide chain of MW 42,000. Under physiologic conditions, actin polymerizes to form double-helical fibers that constitute the core of the thin filament (Fig 15–3). Attached to the actin helix is **tropomyosin**, a rodlike protein similar in structure to the myosin tail. Each tropomyosin molecule interacts with 7 actin monomers in the thin filament. The thin filaments of skeletal and cardiac muscle also contain **troponin**, a protein that contains 3 subunits, designated C, I, and T. The thin filaments of smooth muscle lack troponin.

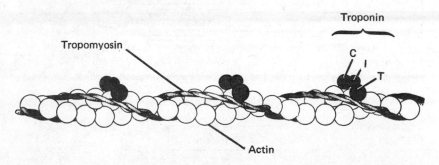

Figure 15–3. The thin filaments of skeletal and cardiac muscle are composed of actin, tropomyosin, and troponin. (Slightly modified and reproduced, with permission, from Martin DW Jr et al: *Harper's Review of Biochemistry,* 20th ed. Lange, 1985.)

Mechanism of Contraction

When a muscle contracts, the sarcomeres become shorter, but the lengths of the thick and thin filaments do not change (Fig 15–4). The sarcomere shortens because the thin filaments slide past the thick filaments toward the center of the sarcomere. In the process, the Z lines are pulled closer together.

Shortening of the sarcomere involves the interaction of actin with myosin and ATP hydrolysis. Each myosin head contains a site at which ATP can be bound and hydrolyzed. In isolation, myosin hydrolyzes ATP very slowly because the products of the reaction, ADP and P_i, remain bound to the active site. However, binding of actin to myosin favors dissociation of ADP and P_i. Once the active site is empty, another molecule of ATP can be bound. Similarly, ATP binding favors dissociation of the actin-myosin complex.

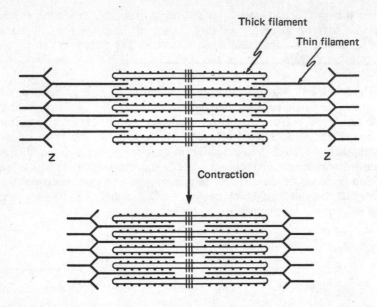

Figure 15–4. During contraction, the thin filaments slide past the thick filaments.

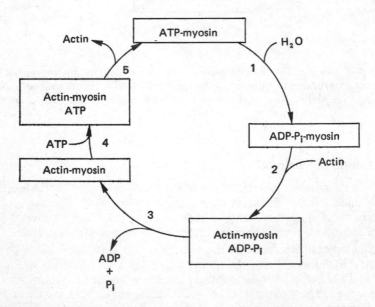

Figure 15–5. The cyclic association and dissociation of actin and myosin is driven by ATP. (Modified and reproduced, with permission, from Stryer L: *Biochemistry,* 2nd ed. Freeman, 1981.)

Subsequent hydrolysis of the ATP allows actin to bind again. Together, these reactions form the cycle summarized in Fig 15–5.

Fig 15–6 illustrates how the cycle of actin-myosin interaction causes the thin filaments to slide relative to the thick ones.

(1) When the myosin head is not attached to the actin filament, it is free to rotate. At that time, it can bind and hydrolyze ATP.

(2) If the actin filament is available, the myosin heads attach to it at a 90-degree angle to the axis of the thin filament.

(3) The myosin molecule then undergoes a conformational change, tilting the myosin head to a 45-degree angle. This pulls the actin filament toward the center of the sarcomere and thus constitutes the power stroke of contraction. Near the end of the power stroke, ADP and P_i dissociate from the myosin head.

(4) Myosin then binds ATP and dissociates from actin. If ATP is not available, a condition known as **rigor,** the myosin cross-bridge remains attached to actin.

Each of the myosin cross-bridges operates independently, and thus the cycle of attachment and detachment of the cross-bridges occurs at different times at adjacent myosin molecules. This allows the muscle to contract with a continuous force. The sarcomere may shorten by 250–625 nm in each contraction, and yet each cross-bridge cycle moves the thin filament only 10 nm. Thus, each contraction involves many rounds of actin-myosin interaction.

Regulation of Contraction

Contraction of a muscle cell is initiated by a signal from the motor neuron that innervates it. Stimulation by the nerve results in depolarization of the plasma membrane of the muscle cell (the **sarcolemma**), which in turn leads to a rise in the concentration of calcium in the sarcoplasm. It is the increase in calcium that is responsible for sarcomere shortening.

In skeletal muscle, the calcium that triggers contraction is released into the sarcoplasm from the **sarcoplasmic reticulum,** a specialized portion of the smooth endoplasmic reticulum that folds around the myofibrils (Fig 15–7). Because the skeletal muscle is large, a signal delivered to its surface would only slowly reach all of the myofibrils in the interior. This problem is solved by a system of transverse tubules (the T tubules) that carry the plasma membrane depolarization into the interior of the cell to the vicinity of the sarcoplasmic reticulum. When the membrane of the sarcoplasmic reticulum is depolarized, the calcium concentration rises from its resting level of 10^{-7} mol/L to 10^{-5} mol/L. The calcium is pumped back into the sarcoplasmic reticulum when the signal from the motor neuron ends. In cardiac muscle cells, contraction is triggered by calcium entering the sarcoplasm both from the sarcoplasmic reticulum and from the extracellular space. Smooth muscle cells have only a rudimentary sarcoplasmic reticulum,

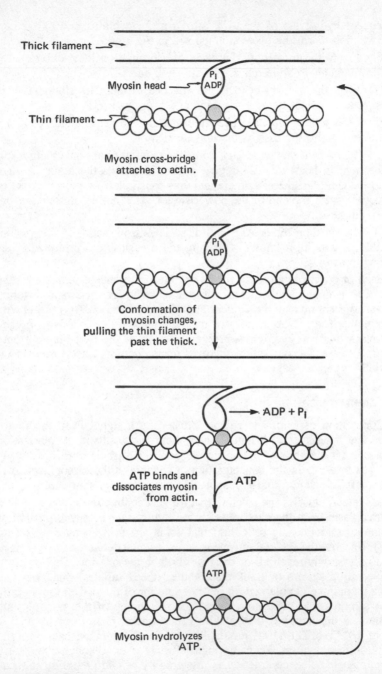

Figure 15–6. The sliding movement of the thick and thin filaments is produced by conformational changes in the myosin head.

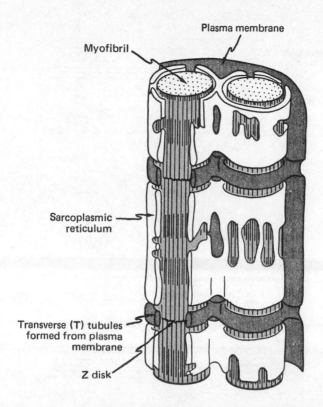

Figure 15–7. In skeletal muscle cells, depolarization of the sarcolemma spreads along the T tubules to the sarcoplasmic reticulum. (Reproduced, with permission, from Alberts B et al: *Molecular Biology of the Cell.* Garland, 1983.)

and most of the calcium that triggers contraction in these cells enters through the sarcolemma.

In skeletal and cardiac muscle cells, the increase in calcium results in contraction because calcium affects the conformation of the thin filament. In the resting cell, the tropomyosin component of the thin filament blocks the access of the myosin cross-bridges to actin and thereby prevents contraction (Fig 15–8). When the concentration of calcium in the cytoplasm increases, calcium binds to the C subunit of troponin. This in turn causes a conformational change in the thin filament that makes the myosin-binding site of actin accessible. While calcium remains bound to troponin, actin and myosin are free to interact and the filaments slide past each other.

Control of contraction in smooth muscle differs from control in skeletal and cardiac muscle in 2 important respects: (1) In smooth muscle, the effect of calcium on the contractile apparatus is mediated not by troponin but by a related calcium-binding protein, **calmodulin.** (2) Smooth muscle myosin

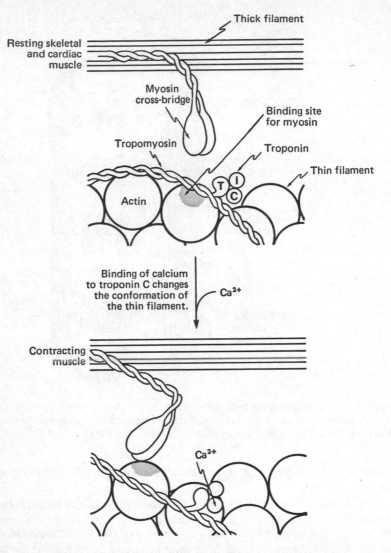

Figure 15–8. Troponin and tropomyosin control the contraction of skeletal and cardiac muscle by regulating the access of myosin to actin. (Redrawn after Katz AM: Congestive heart failure. *N Engl J Med* 1975;**293**:1184.)

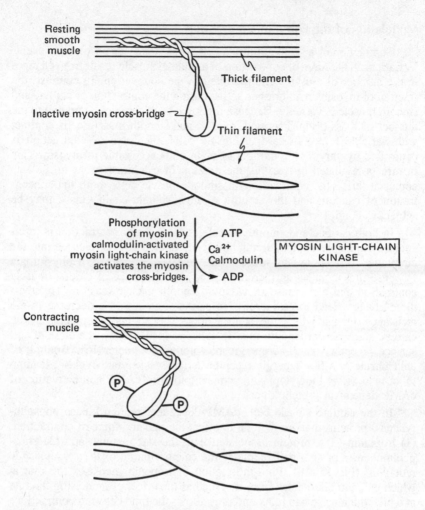

Figure 15–9. Smooth muscle myosin is activated by phosphorylation of the myosin light chains. ⓟ = phosphoryl group.

differs from that of skeletal and cardiac muscle cells in that it can interact with actin only when one of its light chains is phosphorylated. Phosphorylation of myosin is catalyzed by a specialized protein kinase, termed **myosin light-chain kinase** (Fig 15–9). When the concentration of calcium in the smooth muscle cell rises above the resting level, calcium binds to calmodulin, and the calcium-calmodulin complex allosterically activates myosin light-chain kinase. The active kinase then phosphorylates and activates the myosin heads, and the muscle contracts. When the calcium levels fall, the phosphate groups are removed from the myosin light chains by a phosphatase, and the muscle relaxes.

Modulation of the Force of Contraction

Contraction of a skeletal muscle cell is essentially an all-or-none event. Variations in the force of contraction of a skeletal muscle result from changes in the number of muscle cells recruited in the contraction. In contrast, the fraction of myosin cross-bridges active during the contraction of cardiac and smooth muscle is variable. Because the force of contraction is related to the number of cross-bridges formed, cardiac and smooth muscle cells contract with variable force. In cardiac muscle cells, the force of contraction is controlled by varying the number of actin units accessible to myosin. This in turn is regulated by varying the amount of calcium present in the sarcoplasm during the contraction. In smooth muscle cells, both the concentration of calcium and the activity of myosin light-chain kinase may be subject to control.

In both cardiac and smooth muscle cells, force of contraction is influenced by epinephrine, an adrenal hormone. Epinephrine is released into the circulation in response to stress. Once released, it triggers a physiologic response that includes an increase in the rate and force of cardiac muscle contraction and a decrease in vascular smooth muscle tension. Together, these changes lead to an increase in cardiac output and a decrease in the resistance against which the heart works. The effects of epinephrine on cardiac and vascular smooth muscle cells are mediated by one class of cell surface receptor, the β-adrenergic receptor (see Chapter 16). Binding of epinephrine to a β-adrenergic receptor activates adenylate cyclase, leading to an increase in the cytoplasmic concentration of cAMP and activation of cAMP-dependent protein kinase.

In the cardiac muscle cell, cAMP-dependent protein kinase phosphorylates 3 proteins that are thought to affect the rate and force of contraction: (1) troponin; (2) a protein associated with the sarcolemma; and (3) phospholamban, a protein that regulates the calcium pump of the sarcoplasmic reticulum (Fig 15–10). Phosphorylation of troponin increases the rate at which calcium dissociates from it. This has the effect of increasing the rate at which the muscle can relax and thus allows the time between contractions to be shortened. Phosphorylation of the sarcolemma protein is thought to increase the amount of calcium that enters the cell from the extracellular fluid. The increase in calcium results in an increase in the number of actin filaments made accessible and consequently an increase in the number of active myosin cross-bridges. Phosphorylation of phospholamban increases the rate at which the calcium pump of the sarcoplasmic reticulum operates. This has 2 effects on contraction: (1) The rate at which calcium is removed from the cytoplasm is increased, allowing the muscle to relax and contract again more rapidly; and (2) the amount of calcium stored in the sarcoplasmic reticulum is increased, thereby increasing the amount delivered to the sarcoplasm during depolarization.

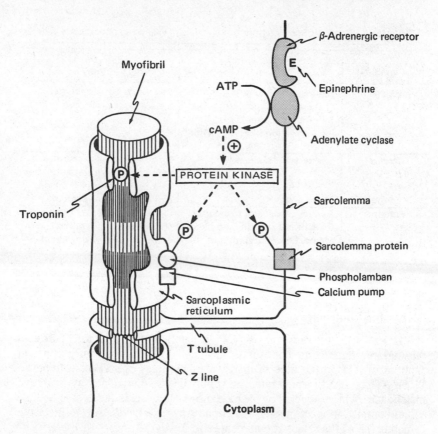

Figure 15–10. Epinephrine increases the force of contraction of cardiac muscle.

In smooth muscle, activation of cAMP-dependent protein kinase results in the phosphorylation and inactivation of myosin light chain kinase. This prevents the activation of the myosin cross-bridges and causes the muscle to relax. Epinephrine may also decrease the entry of calcium into the smooth muscle cell, further decreasing the force of contraction.

The mechanisms by which contraction is controlled and modulated in skeletal, cardiac, and smooth muscle cells are summarized in Table 15–1.

Table 15–1. Control of contraction in skeletal, cardiac, and smooth muscle cells.

Type of Muscle	Regulation of Actin-Myosin Interaction	Effect of Epinephrine
Skeletal	Binding of calcium to troponin alters the conformation of the thin filament, thereby facilitating interaction of myosin with actin.	Stimulates glycogenolysis.
Cardiac	[Same as for skeletal.]	Increases the force and rate of contraction via phosphorylation of troponin, phospholamban, and a sarcolemma protein.
Smooth	Binding of calcium to calmodulin activates myosin light-chain kinase. Phosphorylation of myosin allows it to interact with actin.	Decreases the force of contraction via phosphorylation and inactivation of myosin light-chain kinase.

Energy for Contraction

If a muscle cell runs out of ATP, it enters a state of rigor. To avoid this hazard, muscle metabolism is designed to maintain nearly constant supplies of ATP in the face of highly variable rates of ATP consumption. When ATP demand is moderate, the supply can be maintained by oxidative metabolism, which is fueled by the oxidation of fatty acids and ketone bodies. When contraction is strenuous, these pathways are inadequate to meet the demands for ATP and are supplemented in 2 ways.

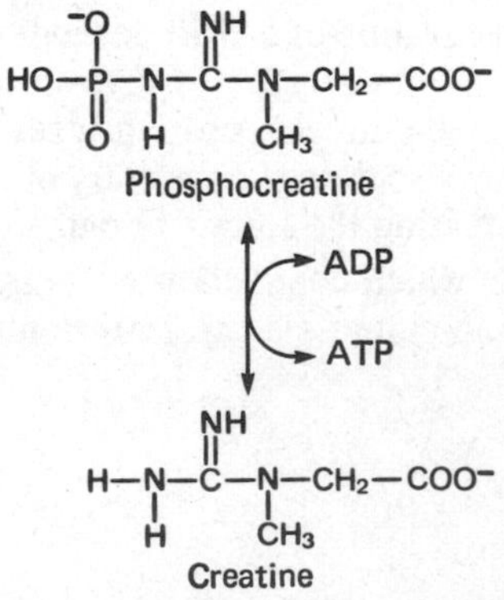

Figure 15–11. Phosphocreatine serves as a reservoir of high-energy phosphate in muscle cells.

(1) If not replenished, the amount of ATP present in skeletal muscle would be consumed in less than 1 second of contraction. To avoid a short-term deficit of ATP, muscle cells store high-energy phosphate in the form of phosphocreatine (Fig 15–11). In the resting muscle, phosphocreatine is formed by the transfer of the terminal phosphate of ATP to creatine. As ATP is consumed in contraction, phosphate is transferred back to ADP.

(2) Skeletal and cardiac muscle can also make use of anaerobic glycolysis to generate ATP. The substrates for glycolysis in muscle come in part from blood glucose supplies. Additional substrates for glycolysis are provided by the breakdown of glycogen, which is triggered during contraction by the increase in the concentration of calcium (see Chapter 8). The effect of calcium on glycogen metabolism is mediated by calmodulin, which constitutes a regulatory subunit of phosphorylase kinase. When the calcium concentration in the sarcoplasm increases, binding of calcium to calmodulin activates phosphorylase kinase, which then activates glycogen phosphorylase and leads to glycogen breakdown.

Tubulin & Dynein

A number of nonmuscle cells, such as sperm, use the whiplike movements of flagella to propel themselves. Other cells, such as those that line

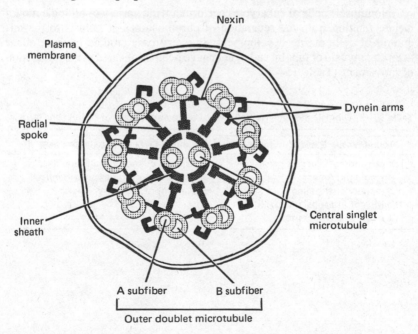

Figure 15–12. The cross-sectional structure of a cilium. (Reproduced, with permission, from Alberts B et al: *Molecular Biology of the Cell.* Garland, 1983.)

the bronchial airways, use cilia to move fluids past their surfaces. Flagella and cilia are similar in structure, and as in the case of muscle, the structure provides a major clue to the mechanism of action.

The movement of cilia is produced by a system of microtubules, shown in cross section in Fig 15–12. The system consists of a central pair of tubules surrounded by an outer circle of 9 double tubules. The tubules themselves are composed of tubulin, a dimeric protein that under physiologic conditions aggregates to form rigid hollow tubes. Attached to each of the outer 9 doublets is a pair of arms composed of another protein, dynein. The dynein arms are spaced along the length of the outer microtubules every 24 nm. The outer doublets are also connected to each other by bands of a highly elastic protein, **nexin,** and they are connected by spokelike structures to a sheath that surrounds the central pair of tubules.

The whiplike motion of cilia results from the sliding of one outer microtubule doublet relative to another. Because the ends of the tubules are fixed at the base of the cilium and because the movement of the tubules is limited by nexin, sliding is converted to a bending motion. The force for sliding is generated by the dynein arms that attach to and detach from the adjacent microtubule. As in muscle, the energy for movement is provided by ATP. If the cilium is depleted of energy, the dynein arms permanently link adjacent microtubules.

Nonmuscle cells of eukaryotes perform a wide variety of biologic movements, ranging from the separation of chromosomes at mitosis to the migration of cells during development. In every case studied to date, either actin and myosin or tubulin and dynein have been implicated in the production of movement (Table 15– 2).

Table 15–2. Biologic movements mediated by actin and myosin or tubulin and dynein.

Actin-Myosin–Based Movement	Tubulin-Dynein–Based Movement
Contraction of muscle	Beating of eukaryotic cilia and flagella
Ameboid movement	Movement of chromosomes in mitosis
Cytoplasmic streaming	Movement of secretory vesicles
Beating of intestinal microvilli	
Cleavage of cells in mitosis	

Hormones | 16

OBJECTIVES

- Be able to name the 4 classes of hormones. Be able to explain how each type of hormone is synthesized and how it exerts its effects.

- Be able to explain the concept of a second messenger.

- Be able to explain how hormones control the synthesis of cAMP.

HORMONES CONSTITUTE a heterogeneous group of chemical messengers that serve to coordinate the activities of different tissues in the body. There are 4 structurally distinct types of hormones: **polypeptide hormones, catecholamines, thyroid hormones,** and **steroids.** This chapter describes the biosynthesis of each of these types of hormone and the general features of hormone action.

A hormone is a chemical messenger that is synthesized in a specific cell type and transported to distantly removed **target cells** via the blood. Most hormones are secreted by endocrine glands, ductless glands that release their products directly into the circulation. Some hormones, however, are secreted by tissues that are not primarily endocrine tissues. Others are secreted by more than one tissue. Norepinephrine, for example, is produced by cells of both the adrenal medulla and the sympathetic nervous system.

A tissue is a target for a given hormone only if the tissue contains specific **receptor proteins** that bind the hormone and initiate a cellular response. Hormones regulate the activities of their target tissues in 2 general ways: (1) by regulating the activities of proteins already present in the cell at the time of hormonal action, and (2) by regulating the synthesis or degradation of proteins. Both types of response change the enzymatic capacity of the cell; the former takes place rapidly (within minutes), while the latter occurs more slowly (requiring a period of hours or days for completion). It is now clear that some hormones use both of these mechanisms.

Hormone receptors can be divided into 2 groups that differ with respect to their cellular location. The receptors for the polypeptide hormones and the catecholamines are displayed on the surface of the cell, whereas those for the steroid and thyroid hormones are inside the cell. Many hormones that bind to cell surface receptors exert their effects on the cell via a **second messenger**—a small molecule whose intracellular concentration is regulated by the hormone and which in turn regulates the activity of one or more enzymes involved in the response to the hormone. Among the molecules that act as second messengers for hormones are calcium, inositol triphosphate, diacylglycerol, cGMP, and cAMP. The intracellular hormone receptors interact directly with chromatin and thereby regulate transcription. Hormones that bind to intracellular receptors are thought not to use second messengers.

Polypeptide Hormones

The polypeptide hormones, ranging in length from as few as 3 to more than 200 amino acids, constitute the largest class of hormones. Although these hormones are produced by a number of tissues and function to control a wide variety of physiologic processes, they share a common mode of synthesis. Because polypeptide hormones are secreted proteins, they are synthesized at the rough endoplasmic reticulum and are matured in the Golgi apparatus. In many cases, proteolytic processing is involved in production of the mature hormone. One representative of this class, insulin, will be discussed here. Insulin, produced by the B cells of the pancreas, is a major regulator of blood glucose levels. High blood glucose levels stimulate the release of insulin from the endocrine pancreas. Insulin, in turn, promotes a decrease in circulating glucose by stimulating its metabolism. The major targets of insulin are adipose tissue, muscle, and liver (see Chapters 8 and 9).

Insulin consists of 2 polypeptide chains, designated A and B, which are linked by 2 disulfide bridges (Fig 16–1). An additional intrachain disulfide bridge forms a loop in the A chain. Insulin is derived from a precursor polypeptide (**preproinsulin**) that contains 2 amino acid sequences not found in the mature hormone: a signal sequence at the N terminus and a connecting peptide, the **C peptide,** located in the primary sequence of the precursor between the A and B sequences (Fig 16–2). Preproinsulin is synthesized by ribosomes attached to the surface of the rough endoplasmic reticulum. As the polypeptide is elongated, it is translocated into the lumen of the rough endoplasmic reticulum, where the signal sequence is removed. This pro-

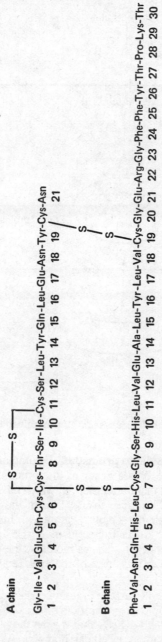

Figure 16–1. The primary structure of insulin. (Reproduced, with permission, from Ganong WF: *Review of Medical Physiology*, 12th ed. Lange, 1985.)

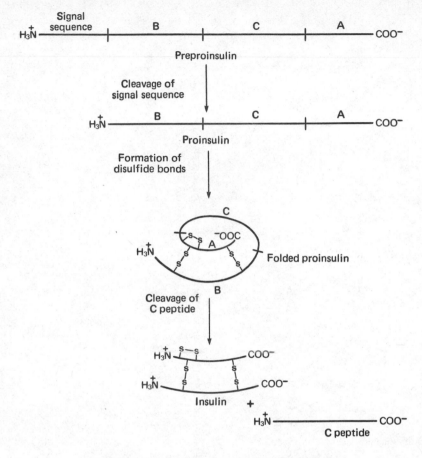

Figure 16–2. Proteolytic processing of preproinsulin.

teolytic step converts preproinsulin to **proinsulin.** Folding of the polypeptide and formation of the disulfide bonds occur in the lumen of the rough endoplasmic reticulum. The final maturation step, in which proinsulin is cleaved to produce insulin and the C peptide, takes place in the Golgi apparatus simultaneously with formation of secretory granules. Insulin and the C peptide are stored until secretion is triggered by a rise in blood glucose levels. In type I (insulin-dependent) diabetes mellitus, the pancreas fails to secrete insulin. Individuals with this form of diabetes require daily insulin replacement.

The insulin receptor is a cell surface protein consisting of 4 chains—2 α chains and 2 β chains—joined by disulfide bridges. The α chains are exposed on the surface of the cell. The β chains span the plasma membrane and have both intracellular and extracellular domains. While the cellular location of the insulin receptor suggests that insulin may act via a second messenger, none of the compounds that have been identified as second messengers for other hormones appear to be involved in insulin action. Recent observations suggest instead that the insulin receptor controls cellular metabolism via covalent modification. The α chains of the receptor form the insulin-binding site and interact with the extracellular domains of the β chains. The intracellular domains contain active sites that have tyrosine-specific protein kinase activity. Binding of insulin to its receptor causes the β chains to phosphorylate themselves and other cellular proteins. It is thought that these phosphorylation reactions lead to the ultimate response of the cell to insulin, which includes both changes in the pattern of gene expression and a decrease in cAMP-dependent phosphorylation of proteins (see Chapters 8 and 9).

Exposure of a target cell to high insulin levels results in a decrease in the number of insulin receptors displayed on its surface and a corresponding decrease in the sensitivity of the cell to insulin. This phenomenon, known as **"down regulation"** of the receptors, results from internalization of the receptor-hormone complex via endocytosis. Some of the internalized complexes are degraded in the lysosomes. Receptor "down regulation" plays a role in type II (non–insulin-dependent) diabetes mellitus. Individuals with this form of diabetes produce insulin; however, the circulating hormone is ineffective in controlling blood glucose levels. Many individuals in this group demonstrate resistance to insulin owing to a decrease in the numbers of insulin receptors on target cells.

Catecholamines

The catecholamines **epinephrine** and **norepinephrine** (Fig 16–3) mediate a physiologic response to emergencies known as the "fight or flight" response. This response includes stimulation of the nervous system, an increase in blood flow to the musculature, and mobilization of fuel supplies. Both epinephrine and norepinephrine are synthesized and released by the adrenal medulla. Norepinephrine is also released by neurons of the sympathetic nervous system and thus functions as both a hormone and a neurotransmitter.

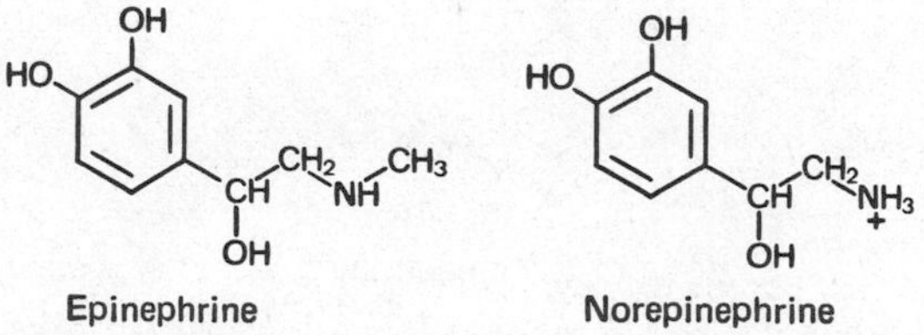

Epinephrine   Norepinephrine

Figure 16–3. The catecholamine hormones.

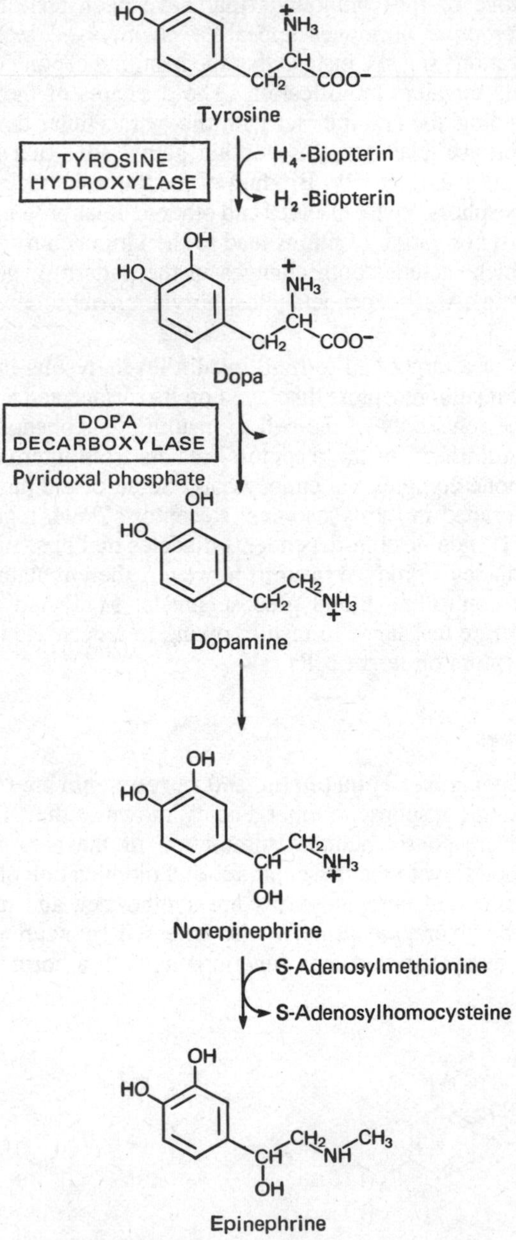

Figure 16-4. Synthesis of the catecholamines.

Fig 16–4 illustrates the pathway by which the catecholamine hormones are synthesized. The first 2 reactions—in which tyrosine is hydroxylated, forming **dopa,** and the latter is decarboxylated, forming **dopamine**—take place in the cytoplasm. Dopamine then enters secretory granules, where it is converted to norepinephrine. In the catecholamine-producing cells of the adrenal medulla, norepinephrine leaves the granules and is converted in the cytoplasm to epinephrine. The latter is then repackaged into secretory granules.

The effects of the catecholamine hormones are mediated by 4 distinct cell surface receptor proteins: the α_1, α_2, β_1, and β_2 **adrenergic receptors.** Epinephrine has a high affinity for both α and β receptors; norepinephrine interacts primarily with α receptors. Many target cells have more than one type of adrenergic receptor, and the response of a given target cell to catecholamines is determined by the number and type of receptors it displays.

Acting through the 4 types of receptors, the catecholamines control the concentrations of 2 second messengers, cAMP and calcium. Stimulation of the α_1-adrenergic receptors leads to an increase in the intracellular concentration of calcium. Stimulation of the β_1 or β_2 receptors results in activation of adenylate cyclase, whereas stimulation of the α_2 receptors has the opposite effect.

The effects of the catecholamines on adenylate cyclase are mediated by 2 GTP-binding proteins, one that stimulates adenylate cyclase (G_s) and one that inhibits it (G_i) (Fig 16–5). When a catecholamine binds to a β-adrenergic

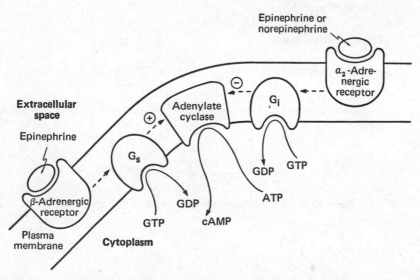

Figure 16–5. Control of adenylate cyclase by G_s and G_i.

receptor, the receptor interacts with G_s, allowing the latter to bind GTP. The G_s-GTP complex activates adenylate cyclase. This effect, however, is short-lived because G_s hydrolyzes the bound GTP and thereby terminates stimulation of adenylate cyclase. The cAMP produced in response to hormone stimulation allosterically activates cAMP-dependent protein kinase, which in turn regulates a number of enzymes by phosphorylation (see Chapters 8, 9, and 15). When a catecholamine binds to an α_2 receptor, G_i is activated, and the activity of adenylate cyclase is depressed. G_s and G_i are not reserved for the adrenergic receptors but also respond to a number of hormones that control cAMP levels, such as glucagon.

Prior to excretion, the catecholamines are catabolized and inactivated by catechol-O-methyltransferase (COMT) and monoamine oxidase (MAO). Fig 16–6 illustrates the reactions involved in the catabolism of epinephrine.

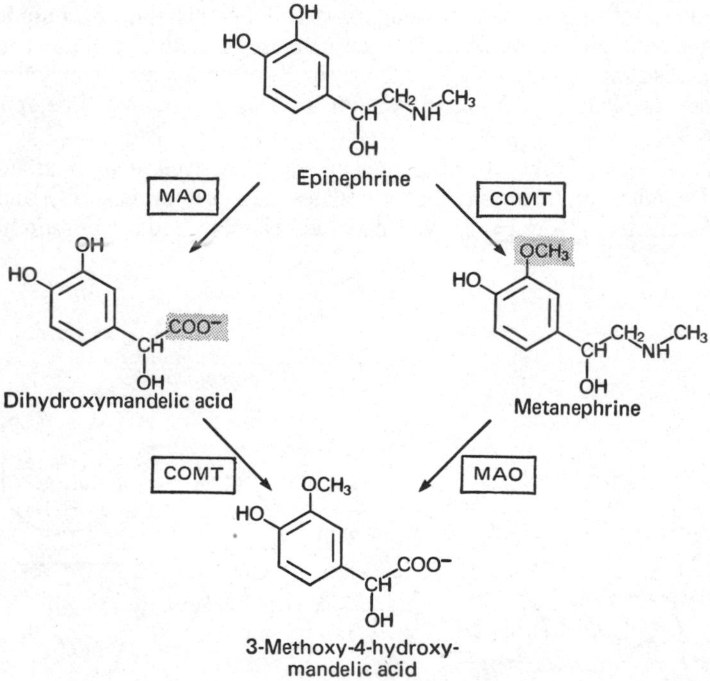

Figure 16–6. Catabolism of epinephrine by COMT and MAO.

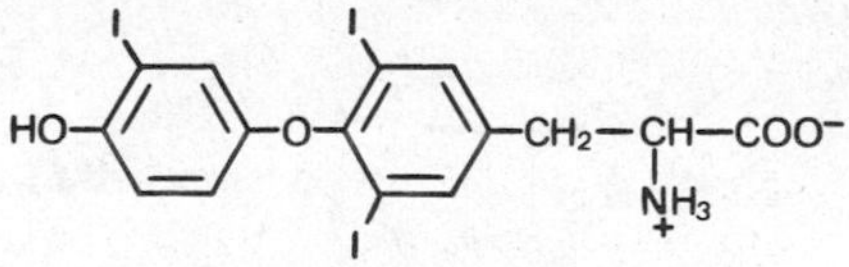

Triiodothyronine (T$_3$)

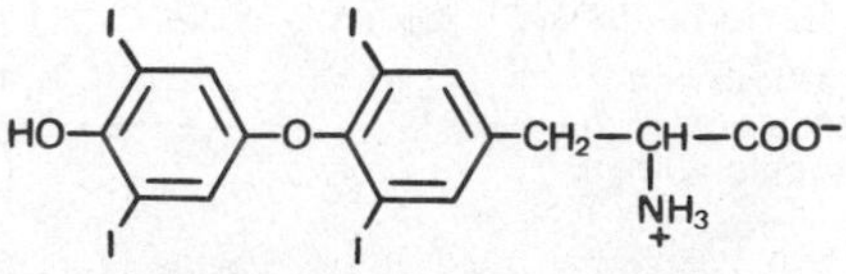

Tetraiodothyronine (T$_4$, thyroxine)

Figure 16–7. The thyroid hormones.

The Thyroid Hormones

The term "thyroid hormone" is applied to 2 iodinated tyrosine derivatives, triiodothyronine (T$_3$) and tetraiodothyronine (T$_4$, thyroxine) (Fig 16–7). T$_3$ and T$_4$ are synthesized by the thyroid gland and act on most tissues to regulate metabolic rate and tissue development. In humans, a lack of thyroid hormone (hypothyroidism) results in impaired growth and mental retardation, while an excess of thyroid hormone (hyperthyroidism) accelerates the metabolic rate.

Synthesis of the thyroid hormones takes place in the follicles of the thyroid gland, each of which consists of a single layer of epithelial cells surrounding a lumen (Fig 16–8). The cells of the follicle synthesize a large, tyrosine-rich precursor protein **(thyroglobulin),** which is secreted into the lumen of the follicle. The cells also concentrate iodide and secrete it into the lumen. The iodide is first oxidized and then added to the tyrosine residues of thyroglobulin, thereby converting them to mono- and diiodothyronine residues. Subsequently, some of the iodinated tyrosine residues are coupled to form T$_3$ and T$_4$ residues. When a follicular cell is stimulated to secrete thyroid hormones, the iodinated precursor is taken up into the cell by endocytosis and transported to the lysosomes. Lysosomal proteases hydrolyze the precursor, yielding T$_3$ and T$_4$, which are then secreted.

The receptors for thyroid hormones are located in the nucleus in association with chromatin. Stimulation of a target cell by thyroid hormone results in changes in gene expression.

The Steroid Hormones

The steroid hormones regulate a wide range of physiologic processes. **Glucocorticoids,** produced in the adrenal cortex, play a role in the regulation of blood glucose and in the response to stress. In high doses, they suppress

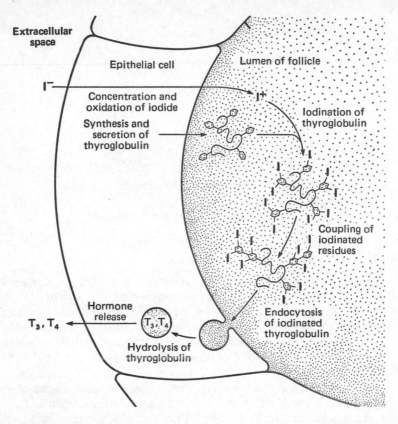

Figure 16–8. Synthesis of thyroid hormones by the follicles of the thyroid gland.

inflammation and the immune response. **Mineralocorticoids,** also produced by the adrenal cortex, participate in the regulation of mineral and water balance by stimulating the reabsorption of Na^+ and Cl^- by the kidney. The resulting increase in osmolality promotes water retention. **Androgens, progesterone,** and **estrogens,** produced by the testes and the ovaries, are required for sexual development.

The glucocorticoids, mineralocorticoids, and progesterone each contain 21 carbon atoms, including a 2-carbon side chain attached to C_{17} (Fig 16–9). Glucocorticoids and mineralocorticoids have a hydroxyl or keto group at C_{11}, while progesterone does not. Because they are structurally similar, the glucocorticoids have some mineralocorticoid activity and vice versa. Androgens and estrogens contain 19 and 18 carbons, respectively. Both lack a C_{17} side chain and instead contain a keto or hydroxyl group at that position. An estrogen can be distinguished from the other steroids by the aromatic form of one of its rings.

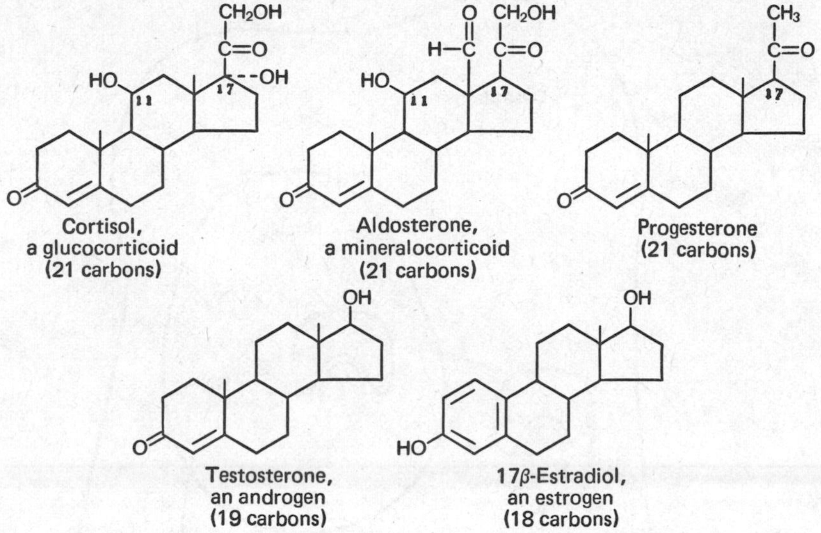

Cortisol,
a glucocorticoid
(21 carbons)

Aldosterone,
a mineralocorticoid
(21 carbons)

Progesterone
(21 carbons)

Testosterone,
an androgen
(19 carbons)

17β-Estradiol,
an estrogen
(18 carbons)

Figure 16–9. Examples of each of the major classes of steroid hormones.

The first and rate-limiting step in the synthesis of steroid hormones involves cleavage of the cholesterol side chain to form pregnenolone (Fig 16–10). The rate of pregnenolone production is largely determined by the availability of cholesterol. The cholesterol consumed in this pathway is provided by intracellular stores of cholesteryl ester and by circulating lipopro-

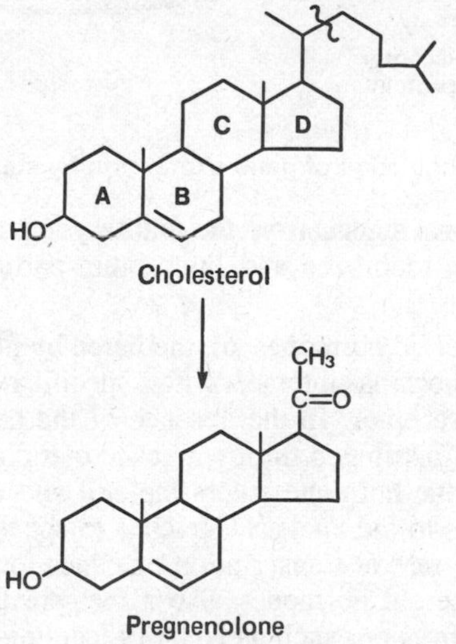

Cholesterol

Pregnenolone

Figure 16–10. The rate-limiting reaction in steroid biosynthesis.

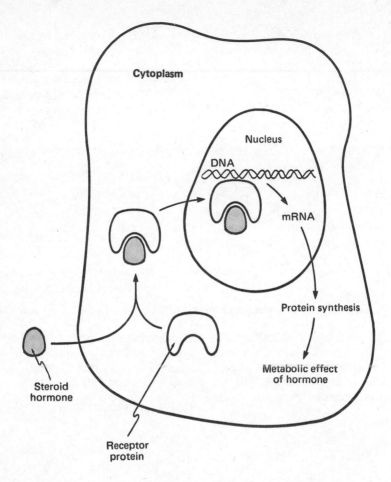

Figure 16–11. Regulation of gene expression by steroid hormones.

tein particles. Hormonal signals from the pituitary regulate the rates at which cholesterol stores are mobilized and lipoprotein particles are taken up by the cell.

The effects of steroid hormones are mediated by intracellular receptors. Each class of steroid hormone interacts with a specific receptor, eg, estrogens bind to an estrogen receptor. In the absence of the homologous hormone, steroid receptors are distributed throughout the cytoplasm of the target cell (Fig 16–1). When the hormone enters the cell and binds to its receptor, the complex migrates to the nucleus, attaches to specific sites on the chromatin, and alters the rate of transcription of genes located near those sites (see Chapter 13). Steroid hormones also affect posttranscriptional events involved in gene expression, such as RNA processing, transport of mRNA to the cytoplasm, and translation.

QUESTIONS: SECTION I

DIRECTIONS (items 1–26): Each numbered item or incomplete statement in this section is followed by answers or by completions of the statement. Select the *one* lettered answer or completion that is *best* in each case.

1. Which of the following amino acids is usually found in the core of a globular protein?
   (A) Ile.
   (B) Asn.
   (C) Glu.
   (D) Ser.
   (E) Thr.

2. If a given histidine residue in myoglobin has a pK of 8.0, what portion of the myoglobin chains will carry a positive charge on this residue at pH 7.0?
   (A) 1%.
   (B) 9%.
   (C) 50%.
   (D) 91%.
   (E) 99%.

3. The graph below documents the effect of caffeine on the enzyme glycogen phosphorylase.

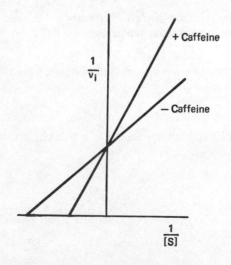

With respect to glycogen phosphorylase, caffeine is—
(A) an allosteric activator.
(B) a noncompetitive inhibitor.
(C) a competitive inhibitor.
(D) none of the above.

4. Which of the following statements about DPG is true?
(A) DPG stabilizes the R conformation of hemoglobin.
(B) The DPG concentration of red blood cells decreases in response to oxygen insufficiency.
(C) DPG binds to hemoglobin but not to myoglobin.
(D) DPG increases the affinity of hemoglobin for oxygen.

5. If the $\Delta G^{o\prime}$ for a reaction is -8.2 kcal/mol at 25 °C, the K_{eq} for this reaction is about—
(A) 10^{-6}.
(B) 10^6.
(C) 10^{-8}.
(D) 10^8.

6. All of the following statements concerning the binding of oxygen to myoglobin are correct *except*—
(A) The His F8 residue of myoglobin reduces the affinity of heme for oxygen.
(B) Oxygen binds to the Fe^{3+} form of heme.
(C) Oxygen binding causes a change in the conformation of the myoglobin polypeptide.
(D) The oxygen saturation curve of myoglobin is hyperbolic in shape.

7. All of the following have a physiologically significant effect on the affinity of hemoglobin for oxygen *except*—
(A) P_{O_2}.
(B) P_{CO_2}.
(C) [Asp].
(D) $[H^+]$.
(E) [DPG].

8. Which of the following amino acids has a side chain that is negatively charged at physiologic pH?
(A) Ala.
(B) Cys.
(C) Tyr.
(D) Asn.
(E) Asp.

9. Carbon monoxide is toxic because it—
   (A) favors a conformation of hemoglobin that has a low affinity for oxygen.
   (B) prevents the binding of DPG to hemoglobin.
   (C) decreases the pH of the red blood cell.
   (D) competes with oxygen for binding to the heme iron of hemoglobin and myoglobin.
   (E) oxidizes heme Fe^{2+} to Fe^{3+}.

10. The free energy change for a pathway that consists of a series of 5 reactions—
    (A) is equal to the sum of the free energy changes of the component reactions.
    (B) is equal to the free energy change for the most favorable reaction of the series.
    (C) is equal to the free energy change for the least favorable reaction of the series.
    (D) cannot be determined from the free energy changes of the component reactions.

11. All of the following statements concerning the active site of an enzyme are correct *except*—
    (A) The active site is generally a cleft or groove in the tertiary structure of the enzyme.
    (B) The amino acids that constitute the active site are contiguous in the primary structure.
    (C) The active site includes the catalytic groups of the enzyme.
    (D) The reaction specificity of an enzyme is determined by groups located within the active site.

12. Methemoglobin—
    (A) contains Fe^{3+} instead of Fe^{2+}.
    (B) binds oxygen irreversibly.
    (C) does not have a quaternary structure.
    (D) is unable to bind carbon dioxide.
    (E) is a mutant form of hemoglobin that has an altered heme binding pocket.

13.

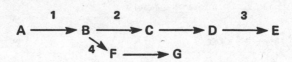

The diagram on the previous page depicts a branched metabolic pathway. A through G represent intermediates in the pathway and 1 through 4 represent enzymes that catalyze several of the reactions of the pathway. Which of the enzymes is most likely to be regulated by the concentration of compound E?

(A) 1.
(B) 2.
(C) 3.
(D) 4.

14. Individuals homozygous for the hemoglobin S gene have an impaired capacity to transport oxygen because–
    (A) the heme iron of hemoglobin S is unusually susceptible to oxidation.
    (B) the hemoglobin S protein is unable to undergo the T to R transition.
    (C) the hemoglobin S polypeptide does not form a quaternary structure.
    (D) the number of red blood cells in circulation is reduced as a result of destruction of sickled cells.

15. Based on the titration curve for alanine shown below, at the point indicated by the arrow, the predominant species will be–

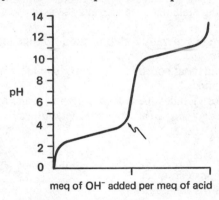

meq of OH⁻ added per meq of acid

(A) $\overset{+}{N}H_3-CH-COOH$
    $\qquad\quad |$
    $\qquad\quad CH_3$

(B) $NH_2-CH-COOH$
    $\qquad\quad |$
    $\qquad\quad CH_3$

(C) $\overset{+}{N}H_3-CH-COO^-$
    $\qquad\quad |$
    $\qquad\quad CH_3$

(D) $NH_2-CH-COO^-$
    $\qquad\quad |$
    $\qquad\quad CH_3$

16. Induced fit refers to–
    (A) the genetic evolution of an active site that closely fits its substrates in shape and charge.
    (B) a change in the structure of the active site of an enzyme that is caused by substrate binding.
    (C) alteration in the structure of a substrate caused by binding to an enzyme.
    (D) conformational change in the active site caused by binding of an effector at another site.

17. A weak acid buffers most effectively–
    (A) at physiologic pH.
    (B) at a pH at which the acid is completely dissociated.
    (C) at a pH at which the acid is completely associated.
    (D) at the pH equal to the pK value for the acid.

18. In an enzyme-catalyzed reaction, the V_{max} is–
    (A) proportionate to the concentration of substrate in the assay.
    (B) equal to twice the K_m.
    (C) independent of the concentration of enzyme that is assayed.
    (D) reached when the enzyme is saturated with substrate.

19. A $\rightleftharpoons$ B; $\Delta G^{o\prime} = +2.7$ kcal/mol

    The above reaction will be at equilibrium when–
    (A) the concentrations of A and B are equal.
    (B) the ratio of B:A is about 3:1.
    (C) the ratio of B:A is about 1:3.
    (D) the ratio of B:A is about 100:1.
    (E) the ratio of B:A is about 1:100.

20. All of the following statements about disulfide bonds are correct *except*–
    (A) Formation of a disulfide bond involves reduction of the cysteine residues.
    (B) A disulfide bond may be formed between 2 cysteine residues that are distant from each other in the primary structure of a protein.
    (C) Disulfide bonds are found in extracellular proteins.
    (D) Two polypeptide chains may be covalently linked by disulfide bonds.

21. Which of the following statements about competitive inhibition is incorrect?
    (A) A competitive inhibitor impairs formation of the ES complex.
    (B) The effect of a competitive inhibitor can be overcome by increasing the concentration of the substrate with which it competes.
    (C) A competitive inhibitor decreases the V_{max} of the reaction.
    (D) A competitive inhibitor increases the apparent K_m of the reaction.

22. Which of the following statements applies to either hemoglobin or myo-globin but not to both?
    (A) Oxygen binding is mediated by a heme prosthetic group.
    (B) The polypeptide chain undergoes a conformational change upon binding oxygen.
    (C) The major secondary structure feature of the protein is the α helix.
    (D) Binding of carbon dioxide decreases the affinity of the protein for oxygen.
    (E) Oxygen binding is competitively inhibited by carbon monoxide.

23. The peptide bond is planar because–
    (A) it is aromatic.
    (B) the partial double bond character of the $C-N$ bond prevents its free rotation.
    (C) the partial charges on the CO and NH groups allow them to participate in H bonds.
    (D) steric hindrance between side chains of adjacent amino acid residues prevents free rotation around the $C-N$ bond.

24. Which of the following statements is correct?
    (A) The $\Delta G^{o\prime}$ value of a reaction can be used to predict the rate at which the reaction will proceed.
    (B) If the $\Delta G^{o\prime}$ value of a reaction is 0, the reaction is at equilibrium.
    (C) A reaction that has a free energy change of -7 kcal/mol is freely reversible.
    (D) A reaction that has a free energy change of -3 kcal/mol will occur spontaneously.

25. An enzyme that displays simple Michaelis-Menten kinetics is assayed with and without a noncompetitive inhibitor. If the data are plotted in a double-reciprocal plot, the effect of the inhibitor–
    (A) is to increase the slope of the line and change the x-intercept.
    (B) is to decrease the slope of the line and change the x-intercept.
    (C) is to increase the slope of the line and change the y-intercept.
    (D) is to decrease the slope of the line and change the y-intercept.
    (E) cannot be predicted from the information given.

26. Enzymes A and B catalyze the same reaction, but their K_m values are 10^{-7} mol/L and 10^{-3} mol/L, respectively. If they have the same V_{max} and are assayed at a substrate concentration of 10^{-3} mol/L, the ratio of the velocity of reaction A to that of reaction B (A:B) is approximately–
    (A) 1:1.
    (B) 2:1.
    (C) 20:1.
    (D) 10,000:1.

DIRECTIONS (items 27–37): For each item in this section, *one* or *more* of the numbered options is/are correct. Select–
  (A) if only 1, 2, and 3 are correct.
  (B) if only 1 and 3 are correct.
  (C) if only 2 and 4 are correct.
  (D) if only 4 is correct.
  (E) if all are correct.

27. Which of the following statements concerning enzymes is/are correct?
  (1) Enzymes can be denatured by heat.
  (2) An enzyme may combine covalently with its substrates.
  (3) The conformation of an enzyme may change upon substrate binding.
  (4) An enzyme may contain both a polypeptide and a nonpolypeptide component.

28. Sickling of erythrocytes in sickle cell anemia is favored by–
  (1) low P_{O_2}.
  (2) low CO_2.
  (3) low pH.
  (4) low DPG.

29. Which of the following solutions have a pH of 7.0?

	pK of Acid	$[A^-] : [HA]$
(1)	5	100:1
(2)	6	1:10
(3)	7	1:1
(4)	8	10:1

30. The tertiary structures of proteins are stabilized by
  (1) hydrogen bonds.
  (2) hydrophobic interactions.
  (3) ionic bonds.
  (4) disulfide bonds.

31. Which of the following statements about the peptide bond is/are correct?
  (1) The carbonyl and imino groups of the peptide bond are polar.
  (2) The peptide bond forms a part of the backbone of a polypeptide.
  (3) The peptide bond has a partial double bond character.
  (4) The groups of the peptide bond can participate in hydrogen bonds.

32. To catalyze a reaction, an enzyme must–
   (1) increase the equilibrium constant of the reaction.
   (2) decrease the activation energy of the reaction.
   (3) have a low K_m.
   (4) bind to its substrates.

33. Which of the following statements about carbon dioxide is/are true?
   (1) It reacts with the N termini of hemoglobin polypeptide chains.
   (2) It competes for the carbon monoxide binding site of hemoglobin.
   (3) It allosterically decreases the binding of oxygen to hemoglobin.
   (4) It stabilizes the R form of hemoglobin.

34. Denaturation of a protein disrupts its–
   (1) biologic activity.
   (2) tertiary structure.
   (3) native conformation.
   (4) peptide bonds.

35. Kinetic data for the reaction catalyzed by aspartate transcarbamoylase are shown below.

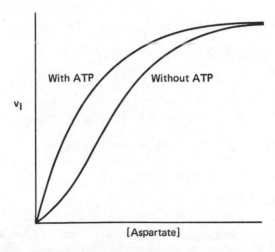

The effect of ATP in this reaction is to–
   (1) increase the V_{max} of the reaction.
   (2) increase the apparent K_m for aspartate.
   (3) decrease the affinity of the enzyme for aspartate.
   (4) abolish the cooperativity of the enzyme for aspartate.

36. Which of the following is/are true of coenzymes?

(1) A coenzyme may constitute a catalytic group in the active site of an enzyme.

(2) Some coenzymes are metabolites of vitamins.

(3) A coenzyme may be covalently attached to the protein with which it functions.

(4) A coenzyme may react covalently with the substrates of the reaction in which it participates.

37. Which of the following statements is/are correct?

(1) An α helix is a secondary structure.

(2) An α helix is stabilized by H bonds between the imino H of one peptide group and the carbonyl O of another.

(3) The side chains of amino acids that form an α helix project to the outside of the helix.

(4) The H bonds of an α helix lie perpendicular to the axis of the helix.

ANSWER KEY: SECTION I

1.	A	2.	D	3.	C
4.	C	5.	B	6.	B
7.	C	8.	E	9.	D
10.	A	11.	B	12.	A
13.	B	14.	D	15.	C
16.	B	17.	D	18.	D
19.	E	20.	A	21.	C
22.	D	23.	B	24.	D
25.	C	26.	B	27.	E
28.	B	29.	B	30.	E
31.	E	32.	C	33.	B
34.	A	35.	D	36.	E
37.	A				

QUESTIONS: SECTION II

DIRECTIONS (items 1–60): Each numbered item or incomplete statement in this section is followed by answers or by completions of the statement. Select the *one* lettered answer or completion that is *best* in each case.

1. Which of the following *cannot* serve as a source of carbon atoms for gluconeogenesis?
   (A) Alanine.
   (B) Pyruvate.
   (C) Lactate.
   (D) Palmitate.
   (E) Oxaloacetate.

2. Which of the following enzymes would be activated during fasting?
   (A) Lipoprotein lipase.
   (B) Hormone-sensitive lipase.
   (C) Acetyl-CoA carboxylase.
   (D) Citrate lyase.
   (E) None of the above.

3. Both acetyl-CoA carboxylase and pyruvate carboxylase–
   (A) utilize thiamin pyrophosphate as a coenzyme.
   (B) are mitochondrial enzymes.
   (C) use ATP as a substrate.
   (D) are activated during fasting.
   (E) are regulated by citrate levels.

4. During fasting, the enzyme responsible for production of free glucose in the liver is–
   (A) glucagon.
   (B) glucose-6-phosphate dehydrogenase.
   (C) glucokinase.
   (D) hexokinase.
   (E) glucose-6-phosphatase.

5. Methotrexate is toxic to human cells because it inhibits–
   (A) reduction of ribonucleotides.
   (B) salvage of purines.
   (C) formation of CTP from UTP.
   (D) reduction of H_2folate to H_4folate.
   (E) synthesis of PRPP.

6. Inhibition of HMG-CoA reductase decreases the rate of synthesis of–
   (A) cholesterol.
   (B) acetoacetate.
   (C) palmitate.
   (D) methylmalonate.
   (E) succinyl-CoA.

7. In humans, a folate cofactor is required for–
   (A) catabolism of purines.
   (B) biosynthesis of homocysteine.
   (C) conversion of dUMP to dTMP
   (D) biosynthesis of uracil.
   (E) reduction of ribonucleotides.

8. Carbon atoms used in the synthesis of fatty acids are transported out of the mitochondria in the form of–
   (A) citrate.
   (B) oxaloacetate.
   (C) malate.
   (D) glycerol phosphate.
   (E) glutamate.

9. Insulin accelerates the–
   (A) production of glucose by the liver.
   (B) uptake of glucose into muscle cells.
   (C) release of fatty acids from adipose tissue.
   (D) conversion of glycogen to glucose in the liver.
   (E) conversion of amino acids to glucose in the muscle.

10. Which of the following statements concerning glucokinase is correct?
    (A) It catalyzes the first step in glycolysis in the brain, muscle, and adipose tissues.
    (B) Its K_m for glucose is below the normal range of blood glucose concentrations.
    (C) In normal individuals, glucokinase plays a role in the stabilization of blood glucose levels.
    (D) In untreated diabetes, glucokinase synthesis is induced.
    (E) Glucokinase is allosterically inhibited by one of its products.

11. Which of the following is true of LDL?
    (A) They participate in the delivery of triglycerides to peripheral tissues.
    (B) Cholesteryl esters constitute a major surface component.
    (C) Apolipoprotein C is required for function.
    (D) The particle is removed from circulation by receptor-mediated endocytosis.

12. Which of the following enzymes plays a role in the Cori cycle?
(A) Lactate dehydrogenase.
(B) Glucose-6-phosphate dehydrogenase.
(C) Pyruvate dehydrogenase.
(D) Glucokinase.
(E) HMG-CoA reductase.

13. Concerning PRPP, all of the following are true *except* –
(A) It is an intermediate in de novo purine biosynthesis.
(B) It is an intermediate in de novo pyrimidine biosynthesis.
(C) It is produced by phosphorylation of ribose 1-pyrophosphate.
(D) It is present in abnormally high concentrations in individuals with Lesch-Nyhan syndrome.

14. Allopurinol is effective in the treatment of hyperuricemia because it–
(A) inhibits xanthine oxidase.
(B) stimulates excretion of uric acid.
(C) inhibits de novo synthesis of purine ribonucleotides.
(D) accelerates purine salvage.
(E) stimulates catabolism of uric acid.

15. Which of the following enzymes of glycolysis catalyzes a thermo-dynamically *reversible* reaction?
(A) Hexokinase.
(B) Pyruvate kinase.
(C) Phosphofructokinase.
(D) Phosphoglycerate kinase.

16. Which of the following is *not* an intermediate in the synthesis of heme?
(A) Coproporphyrinogen III.
(B) Porphobilinogen.
(C) δ-aminolevulinate.
(D) Uroporphyrin III.
(E) Protoporphyrin IX.

17. What is the yield of ATP in the conversion of 1 mol of glucose 6-phosphate to lactate?
(A) 2 mol of ATP.
(B) 3 mol of ATP.
(C) 5 mol of ATP.
(D) 6 mol of ATP.
(E) 8 mol of ATP.

18. Which of the following statements is incorrect?
    (A) Bilirubin is conjugated in the liver and excreted in the bile.
    (B) Conjugation of bilirubin increases its polarity.
    (C) Conjugated bilirubin contains 2 taurine residues.
    (D) Conjugated bilirubin cannot cross the blood-brain barrier.

19. Which one of the following statements concerning the electron transport chain is false?
    (A) The proteins that make up the electron transport chain are associated with the inner mitochondrial membrane.
    (B) The electron transport chain pumps protons from the mitochondrial matrix to the intermembrane space.
    (C) Uncouplers of oxidative phosphorylation inhibit electron transport.
    (D) The electron transport chain accepts reducing equivalents from NADH and $FADH_2$.
    (E) Heme serves as a prosthetic group for the cytochromes of the electron transport chain.

20. Which of the following statements concerning xanthine oxidase is false?
    (A) It converts xanthine to uric acid.
    (B) It forms part of the pathway for the degradation of guanine.
    (C) It is activated by allopurinol.
    (D) It converts hypoxanthine to xanthine.

21. Squalene is an intermediate in the biosynthesis of—
    (A) cholesterol.
    (B) β-hydroxybutyrate.
    (C) methylene H_4folate.
    (D) cytidine triphosphate.
    (E) leucine.

22. Which of the following does not play a role in glucose homeostasis?
    (A) Liver glycogen is broken down to form glucose.
    (B) Pyruvate is converted to glucose in the liver.
    (C) Muscle glycogen is broken down to glucose 6-phosphate.
    (D) Liver glucokinase phosphorylates excess glucose to glucose 6-phosphate.

23. Which of the following statements concerning glucagon is false?
    (A) Its concentration in circulation increases as blood glucose levels decline.
    (B) The liver is a target of glucagon action.
    (C) Glucagon binds to a receptor located on the surface of its target cells.
    (D) Glucagon acts by decreasing the intracellular concentration of cAMP.
    (E) Glucagon is synthesized by the pancreas.

24. Sulfonamides poison bacteria by–
(A) inhibiting synthesis of folic acid.
(B) preventing synthesis of para-aminobenzoic acid (PABA).
(C) preventing salvage of purine nucleosides.
(D) inhibiting synthesis of PRPP.
(E) none of the above.

25. Which of the following is false?
(A) Adenosine deaminase plays a role in both catabolism and salvage of purine nucleosides.
(B) Deficiency of adenosine deaminase causes an increase in dATP pools.
(C) Individuals deficient in adenosine deaminase activity are immuno-deficient.
(D) Deficiency of adenosine deaminase prevents the de novo synthesis of inosine monophosphate (IMP).
(E) Adenosine deaminase converts deoxyadenosine to deoxyinosine.

26. Which of the following statements concerning phenylketonuria (PKU) is incorrect?
(A) In individuals with phenylketonuria, tyrosine hydroxylase is deficient.
(B) For individuals with phenylketonuria, tyrosine is an essential amino acid.
(C) Phenylalanine levels are elevated in individuals with phenylketonuria.
(D) Individuals with phenylketonuria synthesize abnormally large amounts of phenylpyruvate.

27. Which of the following is *not* involved in the esterification of cholesterol by the centripetal transport system?
(A) Lecithin-cholesterol acyltransferase.
(B) Phosphatidylcholine.
(C) Apolipoprotein A.
(D) Apolipoprotein B-100.
(E) HDL.

28. Which of the following is involved in the mobilization of fatty acids from adipose tissue?
(A) Induction of lipoprotein lipase.
(B) Inhibition of triacylglycerol lipase (hormone-sensitive lipase).
(C) Activation of cAMP-dependent protein kinase.
(D) Activation of carnitine acyltransferase.
(E) Activation of acetyl-CoA carboxylase.

29. Which of the following describes the role of VLDL in lipid transport?
(A) Transport of triglycerides from liver to adipose tissue and muscle.
(B) Transport of hepatic cholesterol to peripheral tissues.
(C) Transport of lecithin to adipose tissue and muscle.
(D) Transport of dietary cholesterol to liver.
(E) None of the above.

30. Methotrexate cytotoxicity can be reversed by administering–
(A) thymidine.
(B) N^5-formyl-H_4folate.
(C) folic acid.
(D) hypoxanthine.
(E) dihydrofolate.

31. All of the following statements about fatty acid synthesis in humans are true *except*–
(A) The product of fatty acid synthase is palmitate.
(B) Fatty acid synthesis takes place in the cytoplasm.
(C) The pentose phosphate pathway contributes reducing equivalents.
(D) Oxaloacetate is an intermediate in synthesis of fatty acids from glucose.
(E) Lipogenesis takes place primarily in liver and adipose tissue.

32. Phosphatidylcholine–
(A) is a major storage form for fuel lipids.
(B) is a substrate of lecithin-cholesterol acyltransferase.
(C) includes ceramide within its structure.
(D) is derived from sphingosine.

33. Methotrexate inhibits–
(A) CTP synthase.
(B) ribose-phosphate pyrophosphokinase.
(C) dihydrofolate reductase.
(D) xanthine oxidase.
(E) ribonucleotide reductase.

34. Of the following amino acids, the only one that is strictly ketogenic is–
(A) Tyr.
(B) Trp.
(C) Phe.
(D) Ile.
(E) Leu.

35. Phosphofructokinase—
    (A) is stimulated by citrate and AMP.
    (B) catalyzes the conversion of glucose 6-phosphate to fructose 6-phosphate.
    (C) is a key regulatory enzyme in gluconeogenesis.
    (D) catalyzes an irreversible reaction under physiologic conditions.
    (E) does all of the above.

36. The major regulatory step in the de novo synthesis of cholesterol involves formation of—
    (A) mevalonate.
    (B) HMG-CoA.
    (C) squalene.
    (D) lanosterol.
    (E) none of the above.

37. Which one of the following compounds is a purine nucleoside?
    (A) Uracil.
    (B) Adenine.
    (C) Guanosine monophosphate.
    (D) Thymidine.
    (E) Adenosine.

38. All of the following enzymes are involved in the conversion of pyruvate to glucose *except*—
    (A) pyruvate carboxylase.
    (B) glucose-6-phosphatase.
    (C) phosphoenolpyruvate carboxykinase.
    (D) pyruvate kinase.
    (E) fructose-1,6-bisphosphatase.

39. Each of the following is an intermediate or a product of the urea cycle *except*—
    (A) ornithine.
    (B) arginine.
    (C) citrulline.
    (D) fumarate.
    (E) lysine.

40. Which of the following is *not* involved in the synthesis of fatty acids from glucose?
    (A) Citrate lyase.
    (B) Phosphofructokinase.
    (C) Pyruvate carboxylase.
    (D) Thiamin pyrophosphate.
    (E) Biotin.

41. Which of the following groups of enzymes participates in gluconeo-
genesis but not in glycolysis?
(A) Glucose-6-phosphatase, PEP carboxykinase, acetyl-CoA carboxyl-
ase.
(B) Pyruvate carboxylase, pyruvate kinase, fructose-1,6-bisphospha-
tase.
(C) Pyruvate dehydrogenase, PEP carboxykinase, phosphofructokinase.
(D) Pyruvate carboxylase, PEP carboxykinase, fructose-1,6-bisphos-
phatase.
(E) Glucokinase, fructokinase, acetyl-CoA carboxylase.

42. In which of the following reactions is an H_4folate cofactor oxidized?
(A) dUMP $\rightarrow$ dTMP.
(B) ADP $\rightarrow$ dADP.
(C) Homocysteine $\rightarrow$ methionine.
(D) Methionine $\rightarrow$ S-adenosylmethionine.
(E) None of the above.

43. All of the following statements concerning the de novo synthesis of
purine nucleotides are true *except* –
(A) PRPP is a substrate of the pathway.
(B) Glutamine is the source of 2 of the nitrogen atoms of the purine
ring.
(C) Formation of the N-glycosidic linkage occurs after completion of
the base structure.
(D) Folate cofactors contribute carbons to the purine ring.
(E) Inosine monophosphate is a precursor of both adenosine mono-
phosphate and guanosine monophosphate.

44. All of the following are essential amino acids *except* –
(A) Val.
(B) Pro.
(C) Lys.
(D) Thr.
(E) Met.

45. Triacylglycerol lipase (hormone-sensitive lipase) –
(A) catalyzes the breakdown of triglycerides carried in the core of VLDL.
(B) is induced by insulin.
(C) is regulated by phosphorylation and dephosphorylation.
(D) is an integral part of HDL.

46. In humans, pyruvate is the substrate for several different reactions. Which one of the following reactions would be favored by a high concentration of cytoplasmic NADH?
(A) Pyruvate → oxaloacetate.
(B) Pyruvate → acetyl-CoA.
(C) Pyruvate → lactate.
(D) Pyruvate → alanine.

47. Which of the following reactions of the citric acid cycle is accompanied by a substrate level phosphorylation?
(A) Acetyl-CoA + oxaloacetate → citrate.
(B) Malate → oxaloacetate.
(C) α-Ketoglutarate → succinyl-CoA.
(D) Succinyl-CoA → succinate.
(E) Succinate → fumarate.

48. Ribonucleotide reductase catalyzes the reduction of–
(A) ribose to deoxyribose.
(B) ribonucleosides to deoxyribonucleosides.
(C) ribonucleoside monophosphates to deoxyribonucleoside monophosphates.
(D) ribonucleoside diphosphates to deoxyribonucleoside diphosphates.
(E) ribonucleoside triphosphates to deoxyribonucleoside triphosphates.

49. Deficiency of vitamin B_{12} results in the accumulation of–
(A) dTMP.
(B) purines.
(C) methionine.
(D) methylene-H_4folate.
(E) methylmalonate.

50. In hypoxia (shortage of oxygen), the activity of glycolysis increases because–
(A) cytoplasmic NADH levels are elevated.
(B) lactate is converted to pyruvate.
(C) insulin levels are elevated.
(D) the ratio of AMP:ATP rises.

51. The photosensitivity associated with certain porphyrias is caused by accumulation of–
(A) products of heme catabolism.
(B) heme.
(C) porphyrinogens.
(D) porphobilinogen.
(E) porphyrins.

52. A deficiency of carnitine acyltransferase would impair–
(A) synthesis of nonessential amino acids.
(B) catabolism of fatty acids.
(C) catabolism of glucogenic amino acids.
(D) synthesis of ribonucleotides.
(E) catabolism of keto acids.

53. Which of the following would result from a total deficiency of apolipo-protein B-48?
(A) VLDL would not be formed.
(B) Chylomicrons would not be formed.
(C) LDL would not be formed.
(D) Hepatic cholesterol synthesis would be impaired.

54. Catabolism of fatty acids stimulates gluconeogenesis in all of the following ways *except*–
(A) by contributing to the production of NADH.
(B) by providing carbons for the glucose skeleton.
(C) by activating pyruvate carboxylase.
(D) by contributing to the production of ATP.

55. Which of the following coenzymes serves as a carrier of one-carbon groups?
(A) Pyridoxal phosphate.
(B) Nicotinamide adenine dinucleotide.
(C) H_4folate.
(D) Flavin adenine dinucleotide.
(E) Thiamin pyrophosphate.

56. Which of the following is *not* involved in the oxidation of fatty acids?
(A) Carnitine.
(B) FAD.
(C) Biotin.
(D) CoA.
(E) NAD.

57. Regulation of heme biosynthesis takes place at the step in which–
(A) δ-aminolevulinic acid is formed.
(B) δ-aminolevulinic acid is converted to porphobilinogen.
(C) the linear tetrapyrrole is cyclized.
(D) uroporphyrinogen III is decarboxylated.
(E) iron is inserted into protoporphyrin.

58. Fatty acids that are released from adipose tissue are transported in circulation –
(A) bound to albumin.
(B) as triglycerides carried by chylomicrons.
(C) as lecithin in the surface monolayer of high-density lipoprotein particles.
(D) by none of the above.

59. Which of the following is required for the conversion of heme to bilirubin in vivo?
(A) Glucuronic acid and aminolevulinic acid.
(B) Carbon monoxide and biliverdin.
(C) NADPH and oxygen.
(D) Glucuronic acid and NADH.

60. Which of the following statements concerning the urea cycle is false?
(A) The nitrogen-containing substrates of the urea cycle are ammonia and aspartate.
(B) The cycle consumes ATP.
(C) A nutritionally essential amino acid is synthesized by the cycle.
(D) All of the reactions of the urea cycle take place in the mitochondria.
(E) Fumarate is a product of the cycle.

DIRECTIONS (items 61–83): For each item in this section, *one* or *more* of the numbered options is/are correct. Select–
(A) if only 1, 2, and 3 are correct.
(B) if only 1 and 3 are correct.
(C) if only 2 and 4 are correct.
(D) if only 4 is correct.
(E) if all are correct.

61. Transamination reactions–
(1) are thermodynamically irreversible.
(2) play a role in the synthesis of nonessential amino acids.
(3) are catalyzed by enzymes that require biotin for activity.
(4) play a role in the catabolism of amino acids.

62. Which of the following statements concerning malonyl-CoA is/are true?
(1) It regulates the activity of the carnitine shuttle.
(2) It is an intermediate in fatty acid synthesis.
(3) It is formed in a biotin-dependent reaction.
(4) It is synthesized in the mitochondria.

63. Which of the following properties is/are shared by pyruvate dehydrogenase and α-ketoglutarate dehydrogenase?
    (1) Both are multienzyme complexes.
    (2) Both catalyze a reaction involving CoA.
    (3) Both utilize thiamin pyrophosphate, lipoic acid, and FAD as coenzymes.
    (4) Both are located in the mitochondria.

64. A partial deficiency of uroporphyrinogen cosynthase results in –
    (1) increased production of uroporphyrinogen I.
    (2) induction of δ-aminolevulinate synthase.
    (3) photosensitivity.
    (4) jaundice.

65. The hexose monophosphate shunt is important for the production of –
    (1) glucose.
    (2) pentoses.
    (3) NADH.
    (4) NADPH.

66. Deficiency of the B-100 receptor would lead to –
    (1) elevated levels of circulating LDL.
    (2) decreased lipogenesis in liver.
    (3) increased HMG-CoA reductase activity.
    (4) inhibition of lecithin-cholesterol acyltransferase.

67. Metabolites of vitamin B_{12} play a role in –
    (1) catabolism of fatty acids that contain an odd number of carbon atoms.
    (2) formation of acetyl-CoA from pyruvate.
    (3) transfer of a methyl group from an H_4folate coenzyme to homocysteine.
    (4) synthesis of palmitate.

68. In mammalian cells, carbamoyl phosphate is an intermediate in the biosynthesis of –
    (1) uridine monophosphate.
    (2) inosine monophosphate.
    (3) urea.
    (4) glutamine.

69. The citric acid cycle –
    (1) is involved in the net synthesis of glucose from acetyl-CoA.
    (2) requires pyridoxal phosphate and lipoic acid for activity.
    (3) can operate both aerobically and anaerobically.
    (4) is regulated by ATP and NADH levels.

70. Which of the following statements concerning citrate is/are true?
   (1) It inhibits phosphofructokinase.
   (2) It shuttles 2-carbon units out of the mitochondria.
   (3) It allosterically activates acetyl-CoA carboxylase.
   (4) It inhibits β-oxidation.

71. Conversion of propionyl-CoA to succinyl-CoA requires–
   (1) pyridoxal phosphate.
   (2) biotin.
   (3) thiamin pyrophosphate.
   (4) adenosylcobalamin.

72. HMG-CoA is an intermediate in the synthesis of–
   (1) cholesterol.
   (2) citrate.
   (3) acetoacetate.
   (4) palmitate.

73. A high-energy bond is found in–
   (1) glucose 6-phosphate.
   (2) phosphoenolpyruvate.
   (3) fructose 1,6-bisphosphate.
   (4) acetyl-CoA.

74. De novo ribonucleotide biosynthesis is regulated in part by–
   (1) inhibition of glutamine PRPP amidotransferase by IMP.
   (2) inhibition of the conversion of IMP to AMP by ATP.
   (3) availability of PRPP.
   (4) inhibition of carbamoyl phosphate synthase II by UTP.

75. Fatty acids serve as fuels for–
   (1) liver.
   (2) brain.
   (3) muscle.
   (4) red blood cells.

76. Glycogenolysis in skeletal muscle cells is stimulated by–
   (1) cAMP.
   (2) glucagon.
   (3) calcium.
   (4) insulin.

77. In the Cori cycle, lactate–
   (1) is formed from pyruvate in muscle.
   (2) is converted to pyruvate in liver.
   (3) donates reducing equivalents to NAD.
   (4) accepts reducing equivalents from NADH.

78. Gluconeogenesis is stimulated by–
   (1) acetyl-CoA.
   (2) fructose 2,6-bisphosphate.
   (3) glucagon.
   (4) insulin.

79. Direct precursors for the synthesis of *both* purine and pyrimidine nucleotides include–
   (1) aspartate.
   (2) CO_2.
   (3) glutamine.
   (4) glycine.

80. In the reaction catalyzed by the pyruvate dehydrogenase complex, lipoic acid is directly involved in–
   (1) oxidation of a dithiol.
   (2) reduction of NAD.
   (3) reduction of FAD.
   (4) decarboxylation of pyruvate.

81. Which of the following enzymes is/are required for the synthesis of glycogen from fructose?
   (1) Glucose-6-phosphatase.
   (2) Phosphofructokinase.
   (3) Glycogen phosphorylase.
   (4) Fructose-1,6-bisphosphatase.

82. Which of the following statements about cAMP is/are true?
   (1) It is the second messenger of the hormone glucagon.
   (2) It allosterically regulates the catalytic activity of a protein kinase.
   (3) It competes with caffeine for a phosphodiesterase.
   (4) It stimulates glycogenolysis.

83. Which of the following is/are true of ketogenesis?
   (1) Its products are acetoacetate and β-hydroxybutyrate.
   (2) It is performed in liver mitochondria.
   (3) HMG-CoA is an intermediate.
   (4) The rate of ketogenesis increases during fasting.

ANSWER KEY: SECTION II

1.	D	2.	B	3.	C
4.	E	5.	D	6.	A
7.	C	8.	A	9.	B
10.	C	11.	D	12.	A
13.	C	14.	A	15.	D
16.	D	17.	B	18.	C
19.	C	20.	C	21.	A
22.	C	23.	D	24.	A
25.	D	26.	A	27.	D
28.	C	29.	A	30.	B
31.	D	32.	B	33.	C
34.	E	35.	D	36.	A
37.	E	38.	D	39.	E
40.	C	41.	D	42.	A
43.	C	44.	B	45.	C
46.	C	47.	D	48.	D
49.	E	50.	D	51.	E
52.	B	53.	B	54.	B
55.	C	56.	C	57.	A
58.	A	59.	C	60.	D
61.	C	62.	A	63.	E
64.	A	65.	C	66.	B
67.	B	68.	B	69.	D
70.	A	71.	C	72.	B
73.	C	74.	E	75.	B
76.	B	77.	A	78.	B
79.	A	80.	B	81.	D
82.	E	83.	E		

QUESTIONS: SECTION III

DIRECTIONS (items 1–25): Each numbered item or incomplete statement in this section is followed by answers or by completions of the statement. Select the *one* lettered answer or completion that is *best* in each case.

1. Which of the following statements about double-stranded DNA is false?
   (A) The amount of A + G equals the amount of T + C.
   (B) Each of the 2 strands has the same nucleotide sequence.
   (C) The 5' end of one strand is paired with the 3' end of the other strand.
   (D) Both base stacking and base pairing contribute to the stability of the double helix.

2. Which of the following is *not* involved in steroid hormone action?
   (A) Binding of the hormone to an intracellular receptor.
   (B) Production of a second messenger.
   (C) Binding of a receptor-hormone complex to chromatin.
   (D) Regulation of transcription.

3. All of the following are true of smooth muscle *except*–
   (A) Contraction occurs when the myosin light chains are phosphorylated.
   (B) Myosin light-chain kinase is activated by a calcium-calmodulin complex.
   (C) The thin filaments lack troponin.
   (D) Phosphorylation of myosin light-chain kinase increases the force of contraction.

4. In gene expression, the promoter is–
   (A) a DNA sequence at which transcription is initiated.
   (B) the 5' untranslated portion of an mRNA.
   (C) a DNA sequence to which a positive regulatory protein binds.
   (D) an RNA sequence that directs ribosome binding.
   (E) a DNA sequence at which replication is initiated.

5. Which of the following is true concerning membrane lipids?
   (A) Biologic membranes contain phospholipids, triglycerides, and cholesterol.
   (B) Glycolipids are located only on the cytoplasmic face of the membrane.
   (C) Membrane fluidity increases as the ratio of saturated to unsaturated fatty acyl residues increases.
   (D) Lipids are free to move laterally within the plane of the membrane.

6. All of the following statements concerning asparagine-linked oligosaccharides are true *except—*
(A) They are attached to membrane and secreted proteins.
(B) Dolichol serves as a membrane-bound anchor on which the oligosaccharide is assembled.
(C) Their synthesis involves nucleotide-activated sugars.
(D) Attachment of the oligosaccharide to the asparagine residue occurs in the Golgi apparatus.
(E) The oligosaccharide structure may be modified following transfer to the protein.

7. As a result of wobble pairing—
(A) tRNA charging can be proofread by aminoacyl-tRNA synthetases.
(B) a tRNA may be charged with several similar amino acids.
(C) fewer than 61 tRNAs are required to translate mRNA.
(D) the genetic code is ambiguous.

8. Which of the following statements concerning the cap structure of a eukaryotic mRNA is false?
(A) It plays a role in the initiation of translation.
(B) It is located at the 5' end of the polymer.
(C) Its structure includes 7-methylguanosine.
(D) Its structure may include 2' methyl groups.
(E) It is added to the primary transcript by RNA polymerase II.

9. All of the following are true of initiation of translation in eukaryotes *except—*
(A) ATP is hydrolyzed.
(B) Met-tRNA$_i$ binds to the P site of the ribosome.
(C) Selection of the start codon involves base pairing between mRNA and rRNA.
(D) The small ribosomal subunit binds to the mRNA prior to the large subunit.

10. The hormone shown below is—

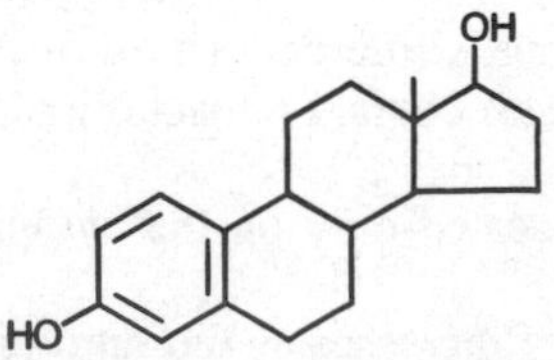

(A) an androgen.
(B) an estrogen.
(C) a glucocorticoid.
(D) a mineralocorticoid.
(E) progesterone.

11. Which of the following does *not* play a role in the replication of a bacterial chromosome?
    (A) DNA helicase.
    (B) UTP.
    (C) DNA ligase.
    (D) DNA gyrase.
    (E) Reverse transcriptase.

12. The sigma (σ) subunit of *Escherichia coli* RNA polymerase–
    (A) acts only at the termination of transcription.
    (B) requires GTP for activity.
    (C) is involved in the recognition of promoter sequences.
    (D) plays a role in proofreading.
    (E) is the subunit to which substrate nucleotides are bound.

13. Which of the following statements about tRNA is false?
    (A) Each tRNA can be charged with only one species of amino acid.
    (B) Transfer RNAs contain nucleotides other than A, G, C, and U.
    (C) All tRNAs share a common secondary structure.
    (D) The anticodon is at the 5′ end of the molecule.

14. Which of the following statements concerning a eukaryotic mRNA is false?
    (A) It is monocistronic.
    (B) Ribosome binding takes place at the 5′ end.
    (C) The coding sequence may be interrupted by introns.
    (D) It is translated in the cytoplasm.
    (E) The codon corresponding to the N terminus of the protein is near the 5′ end.

15. All of the following take place during the elongation steps of protein synthesis *except*–
    (A) An amino acid is transferred from the 3′ end of a tRNA to the amino group of another amino acid.
    (B) Peptidyl-tRNA is transferred from the P site of the ribosome to the A site.
    (C) Two GTP are hydrolyzed for every amino acid incorporated into the polypeptide chain.
    (D) The ribosome moves on the mRNA in the 5′ to 3′ direction.

16. In skeletal muscle cells, calcium triggers glycogen breakdown by binding to –
   (A) troponin.
   (B) calmodulin.
   (C) adenylate cyclase.
   (D) glycogen phosphorylase.

17. Which of the following does *not* play a role in triggering contraction of skeletal muscle?
   (A) Binding of calcium to troponin.
   (B) Phosphorylation of myosin by myosin light-chain kinase.
   (C) Depolarization of the sarcolemma.
   (D) Release of calcium from the sarcoplasmic reticulum.

18. In the termination of protein synthesis –
   (A) the termination codon is read by one of 3 nonsense tRNAs.
   (B) the termination codon is recognized by base pairing between the mRNA and an rRNA.
   (C) a termination factor hydrolyzes the last aminoacyl-tRNA bond.
   (D) peptidyl transferase catalyzes the release of the completed polypeptide chain.
   (E) mRNA is degraded.

19. Clathrin –
   (A) is involved in the synthesis of asparagine-linked oligosaccharides.
   (B) is the major protein in the coat of coated vesicles.
   (C) is the anion channel of the red blood cell plasma membrane.
   (D) is an Na^+/Ca^{2+} antiport.

20. All of the following events take place during DNA replication *except* –
   (A) An ATP-requiring enzyme unwinds the parent DNA double helix.
   (B) An RNA polymerase synthesizes primers for DNA polymerase.
   (C) A proofreading exonuclease activity removes unpaired nucleotides from the $5'$ end of the growing chain.
   (D) RNA primers are removed and replaced by DNA.
   (E) DNA ligase joins adjacent Okazaki fragments.

21. Which of the following proteins is an ATPase?
   (A) Myosin.
   (B) Actin.
   (C) Troponin.
   (D) Calmodulin.
   (E) Tropomyosin.

22. In bacterial protein synthesis, elongation factor G–
(A) catalyzes the aminoacylation of tRNAs.
(B) reads the codon displayed in the A site of the ribosome.
(C) facilitates the binding of charged tRNAs to the A site of the ribosome.
(D) catalyzes peptide bond formation.
(E) is involved in the release of uncharged tRNAs from the P site.

23. All of the following statements concerning catecholamines are correct *except*–
(A) Dopamine is an intermediate in their synthesis.
(B) Their effects are mediated by cell-surface receptors.
(C) One hormone can use 2 different second messengers.
(D) Monoamine oxidase catalyzes the rate-limiting step in their synthesis.
(E) They are produced in both the adrenal medulla and neurons of the sympathetic nervous system.

24. Which of the following best describes the role of ATP in muscle contraction?
(A) Binding of ATP to actin regulates the conformation of the thin filament.
(B) Hydrolysis of ATP by myosin shortens the thick filament.
(C) Binding of ATP to myosin favors the dissociation of the actin-myosin complex.
(D) Hydrolysis of ATP by the thin filament favors association of actin with myosin.

25. All of the following statements concerning synthesis of a transmembrane protein of the plasma membrane are true *except*–
(A) Insertion of the protein into a lipid bilayer occurs cotranslationally.
(B) Polymerization is catalyzed by ribosomes bound to the cytoplasmic face of the plasma membrane.
(C) A hydrophobic signal sequence directs translocation of the protein through the membrane.
(D) Glycosylation of the protein occurs as it is inserted into the membrane.
(E) Following translocation through the membrane, the protein may be proteolytically modified.

DIRECTIONS (items 26–35): For each item in this section, *one* or *more* of the numbered options is/are correct. Select–
  (A) if only 1, 2, and 3 are correct.
  (B) if only 1 and 3 are correct.
  (C) if only 2 and 4 are correct.
  (D) if only 4 is correct.
  (E) if all are correct.

26. The insulin receptor–
  (1) is synthesized on ribosomes that attach to the rough endoplasmic reticulum.
  (2) may be internalized by endocytosis.
  (3) is a tyrosine-specific protein kinase.
  (4) regulates the activity of adenylate cyclase.

27. In a tRNA molecule, the sequence CCA is–
  (1) base-paired with a UGG sequence.
  (2) at the 5′ terminus.
  (3) involved in recognition of the anticodon.
  (4) the site of amino acid attachment.

28. Substitution of one nucleotide for another in the coding region of a gene *may*–
  (1) have no effect on the encoded protein.
  (2) result in the substitution of one amino acid for another in the sequence of the protein.
  (3) result in premature termination of the polypeptide chain.
  (4) cause a frameshift during translation.

29. Which of the following statements about Na^+/K^+-ATPase is/are true?
  (1) It is a plasma membrane protein.
  (2) It is inhibited by digitalis.
  (3) It is phosphorylated by ATP.
  (4) It functions to transport sodium into the cell.

30. Membrane proteins may–
  (1) extend through the lipid bilayer.
  (2) diffuse laterally within the plane of the bilayer.
  (3) bind to extracellular proteins.
  (4) transport polar substances across the membrane.

31. Which of the following is/are true of the *lac* repressor?
   (1) It is an allosteric protein whose conformation is regulated by the *lac* inducer.
   (2) It is the product of a structural gene that is expressed constitutively.
   (3) When bound to the *lac* operator, it prevents binding of RNA polymerase to the *lac* promoter.
   (4) Its synthesis is regulated by cAMP.

32. A DNA molecule contains the following sequence:

$$5'. . .CGTTAT. . .3'$$

Which of the following is/are true?
   (1) A transcript of this strand contains the sequence 5′. . .UAUUGC. . .3′.
   (2) The complementary strand of DNA has the sequence 5′. . .GCAATA. . .3′.
   (3) Replication of this strand is from left to right.
   (4) Transcription of this strand is from right to left.

33. In the genetic code−
   (1) many amino acids are encoded by more than one codon.
   (2) each codon specifies only one amino acid.
   (3) one codon specifies initiation of translation.
   (4) 3 codons specify termination of translation.

34. Which of the following statements about poly (A) polymerase is/are true?
   (1) It modifies the 5′ ends of some primary transcripts.
   (2) It is located in the cytoplasm.
   (3) It requires a DNA template.
   (4) It adds 100−200 AMP residues to the end of an RNA chain.

35. Histones−
   (1) are the major proteins of eukaryotic chromatin.
   (2) contain a high percentage of basic amino acids.
   (3) are highly conserved throughout evolution.
   (4) are not found in bacteria.

ANSWER KEY: SECTION III

1.	B	2.	B	3.	D
4.	A	5.	D	6.	D
7.	C	8.	E	9.	C
10.	B	11.	E	12.	C
13.	D	14.	C	15.	B
16.	B	17.	B	18.	D
19.	B	20.	C	21.	A
22.	E	23.	D	24.	C
25.	B	26.	A	27.	D
28.	A	29.	A	30.	E
31.	A	32.	D	33.	E
34.	D	35.	E		

Index